Prometheus and Gaia

Prometheus and Gaia

Technology, Ecology, and Anti-Humanism

Harrison Fluss
and
Landon Frim

ANTHEM PRESS

Anthem Press
An imprint of Wimbledon Publishing Company
www.anthempress.com

This edition first published in UK and USA 2023
by ANTHEM PRESS
75–76 Blackfriars Road, London SE1 8HA, UK
or PO Box 9779, London SW19 7ZG, UK
and
244 Madison Ave #116, New York, NY 10016, USA

First published in the UK and USA by Anthem Press in 2022

Copyright © Harrison Fluss and Landon Frim 2023

The authors assert the moral rights to be identified as the authors of this work.

All rights reserved. Without limiting the rights under copyright reserved above, no part of this publication may be reproduced, stored or introduced into a retrieval system, or transmitted, in any form or by any means (electronic, mechanical, photocopying, recording or otherwise), without the prior written permission of both the copyright owner and the above publisher of this book.

British Library Cataloguing-in-Publication Data
A catalogue record for this book is available from the British Library.

Library of Congress Control Number: 2023935640

ISBN-13: 978-1-83998-926-1 (Pbk)
ISBN-10: 1-83998-926-2 (Pbk)

Cover credit: Prometheus sculpture on the background of Ahun mountain, Sochi, Russia. By FromMyEyes/Shutterstock.com

This title is also available as an e-book.

For Samantha and Lisa

CONTENTS

Acknowledgments ix

Introduction: Fear and Loathing in the Twenty-First Century 1

1. Prometheus and Gaia 23

2. Accelerationism 73

3. Eco-Pessimism 103

4. Coincidence of Opposites 151

Epilogue: Beyond the Void 167

Index 179

ACKNOWLEDGMENTS

We would like to express our thanks to Peter Bloom who, as series editor, gave consistent support and constructive feedback for this project. It is owing to his enthusiasm, and willingness to entertain our criticisms of fashionable ideas, that this book ever saw the light of day. We would also like to acknowledge the editorial team at Anthem Press for their meticulous review of the original manuscript.

We are grateful to all those who read earlier versions of this work, provided critical comments, and were in conversation about its central themes and arguments. These include Benjamin Noys, Ishay Landa, Carolyn Culbertson, Doug Henwood, Liza Featherstone, Doug Greene, Ben Hillin, Zachary Kimes, and Hamad Al-Rayes.

Finally, we wish to acknowledge the invaluable contributions of Daniel Coutu, Coralie Paschal, and Andrew Jepsen who reviewed this book manuscript as part of their philosophical education at Florida Gulf Coast University.

Introduction

FEAR AND LOATHING IN THE TWENTY-FIRST CENTURY

Climate change has inspired new forms of misanthropy. The moderate complaint that humans should "reduce, reuse, and recycle," has given way to a supposedly radical critique that "it's already too late."[1] The environment, this new story goes, is beyond repair in terms of human habitability; Mother Earth is rejecting us as we slide into the "post-Anthropocene." This is what we deserve, after all, for our centuries of ecological degradation. Humanity is either doomed, or at least, needs to be "right-sized" through massive depopulation efforts. Humility and resignation are the order of the day.

Many on the ecological Left, while not always going this far, nonetheless stress the virtue of humility. In particular, John Bellamy Foster from *Monthly Review* diagnoses our current predicament as having deep roots in Enlightenment thinking. The seventeenth-century idea of a "lawful" nature, à la Francis Bacon, is criticized as just a prelude to the arrogant domination of the environment through science and technology.[2]

At the same time, Foster equivocates: Modern capitalism is responsible for opening a violent "ecological rift" precisely by imagining that there are no stable, intelligible limits to nature—that with enough technical know-how, we can reorder existence to our liking. For similar reasons, he derides the eco-modernism of other figures on the Left as promoting a "New Promethean Socialism."[3] Their love of centralized technocracy and titanic projects of geo-engineering are anathema to Foster's degrowth

---

1. Jonathan Franzen, "What If We Stopped Pretending?," *New Yorker*, September 8, 2019, https://www.newyorker.com/culture/cultural-comment/what-if-we-stopped-pretending; Nafeez Ahmed, "The Collapse of Civilization May Have Already Begun," *Vice*, November 22, 2019, https://www.vice.com/en_us/article/8xwygg/the-collapse-of-civilization-may-have-already-begun.
2. John Bellamy Foster, "The Long Ecological Revolution," *Monthly Review: An Independent Socialist Magazine*, November 1, 2017, https://monthlyreview.org/2017/11/01/the-long-ecological-revolution.
3. Foster, "The Long Ecological Revolution."

politics.[4] The solution, it seems, is to reject the values of endless growth and innovation, in favor of a slowed-down society. Of course, opines Foster, capitalism is constitutionally incapable of this kind of value-shift.[5]

In the end, however, it's not just capitalism that is the culprit, but the very ideas of human exceptionalism, autonomy, and control. We should reject any project of humanity becoming the "collective sovereign of [the] Earth" outright.[6]

While not having an identical perspective on these questions, Andreas Malm likewise derides "techno-utopian perspectives" as "completely juvenile and out of touch with material realities." His own brand of ecological Marxism invokes the spirit of "War Communism" as necessary for tackling today's environmental questions.[7] That is, the austerity, sacrifice, and enforced regimentation which marked the Russian Civil War is needed to meet contemporary crises. It is this idea of war and "catastrophe" that inspires Malm's pessimism about any easy technological solutions; sacrifices must be made.

Jason Moore, in his book *Capitalism in the Web of Life* (2015), takes industrial society to task for considering "Nature" as merely an external, objective thing. Capitalism treats the environment as an object to be "coded, quantified, and rationalized" for the ends of "economic growth" and "social development."[8] Again, Moore identifies an extended intellectual tradition as partly culpable for such hubris, this time going back to the mechanistic philosophy of René Descartes. It is at Descartes's feet that Moore also lays the blame for the modern separation between human society and the rest of nature.[9]

---

4. Leigh Phillips, the author of *Austerity Ecology & the Collapse-Porn Addicts* (2015), is the main ideological villain of Foster's polemic. The antipathy appears to be mutual. But while one can appreciate Phillips's insights regarding the environmentalism of the "small is beautiful" crowd, his own admitted anthropocentrism verges on an illicit "speciesism" which arbitrarily places humanity above all other sentient beings. See Leigh Phillips, *Austerity Ecology & the Collapse-Porn Addicts: A Defense of Growth, Progress, Industry and Stuff* (Winchester, UK: Zero Books, 2015), 75–79.
5. John Bellamy Foster, "Capitalism and Degrowth: An Impossibility Theorem," *Monthly Review: An Independent Socialist Magazine*, January 1, 2011, https://monthlyreview.org/2011/01/01/capitalism-and-degrowth-an-impossibility-theorem/.
6. This is Foster's critique of Leigh Phillips and Michal Rozworski's modernist position as described in their 2017 *Jacobin* article. See Leigh Phillips and Michal Rozworski, "Planning the Good Anthropocene," *Jacobin*, August 15, 2017, https://www.jacobinmag.com/2017/08/planning-the-good-anthropocene.
7. Andreas Malm, "To Halt Climate Change, We Need an Ecological Leninism," *Jacobin*, June 15, 2020, https://jacobinmag.com/2020/06/andreas-malm-coronavirus-covid-climate-change.
8. Jason W. Moore, *Capitalism in the Web of Life: Ecology and the Accumulation of Capital* (London: Verso, 2015), 2.
9. Moore, *Capitalism in the Web of Life*, 19.

These writers all share a denial that we can "outgrow" our ecological challenges. Sacrifice is held up as a political ideal. Other authors (including the main subjects of this book) go further still. Self-described "Gaians," including Bruno Latour, Isabelle Stengers, Donna Haraway, Déborah Danowski, and Eduardo Viveiros de Castro, explicitly deny any distinction between "Nature" and "culture," and so construct an explicitly animist worldview.

Such eco-pessimism finds popular expression in contemporary novels, films, and music. One need only recall Pixar's *WALL-E* (2008), Darren Aronofsky's apocalyptic features *Noah* (2014) and *Mother!* (2017), and the anime movie *Weathering with You* (2019). In each of these morality tales, an ecocidal humanity suffers a just, if devastating, punishment.

If we are to ward off disaster, we have to apply the brakes to progress, even if this may have jarring effects. For some environmental activists, one solution is to move toward a "locavorist" model, wherein we consume only what is produced in our immediate region. This "small is beautiful" approach mandates that communal farms replace large corporate entities, substituting diverse, nonintensive methods for intensive, monocrop farming. "Quality is better than quantity," with organic and heirloom products celebrated over cheaply produced calories for mass consumption. This would also cut out the need for complex shipping and supply chains (a major factor in our expanding carbon footprint).

The logical implication of such locavorism, however, is the dismantling of large, urban centers. Many of these (e.g., Chicago, New York, London) have harsh winters inhospitable to year-round crop production. Without complex, global supply chains, New Yorkers will be unable to consume such basic foodstuffs as bananas, rice, or oranges, and the city itself will accommodate only a fraction of its current population. But perhaps this is the point: Restriction of our needs and desires is a good thing, a signal that we are finally transcending our self-destructive egoism, materialism, and consumerism.

At the same time, there appears to be an opposite trend in contemporary culture. For every eco-pessimist and prophet of doom, one finds an ecstatic technophile ready to herald the next era of innovation and progress.[10] If we cannot avoid ecological disaster through small and humble approaches, then we need to go big. As Leigh Phillips energetically put it, "Retreat from our predicament is not an option. We must push through the Anthropocene, indeed *accelerate* our modernity, and accept our species' dominion over the Earth."[11]

---

10. For a popular science book on these contrasting perspectives, see Charles C. Mann, *The Wizard and the Prophet: Two Remarkable Scientists and Their Dueling Visions to Shape Tomorrow's World* (New York: Knopf, 2018).
11. Phillips, *Austerity Ecology*, 186. Unless otherwise noted all italics are in original.

The vague term "Futurist" is often applied to this ideology, especially when it takes a more anti-humanist turn than what Phillips presents. Enthralled by the capacity of new technologies to disclose meaning and significance, Futurists locate the power to create new values within the hands of the inventor, the entrepreneur, and sometimes even the venture capitalist.[12] Conversely, the concept "humanity" is deemed too parochial, small, and fragile to withstand such rapid innovation. For the average, "unenlightened" human is too rooted in outmoded and sentimental ideas of welfare, comfort, attachment, and need.

If *Homo sapiens* cannot keep up with the rapid pace of automation and change, so much the worse for us. This thinking has led many in the Futurist camp to embrace the concept of universal basic income (UBI).[13] If technologies make certain kinds of work redundant, then the solution is to simply pay people to stay home. Such proposals, however well-intentioned, foreground production before workers themselves. The point is to demobilize workers who have been made obsolete by more efficient technology. But what UBI proposals consistently leave out is any notion of worker control over production. Thus, what is often touted by American politicians as a "freedom dividend" is, in fact, a way to expel the mass of workers from the production process, and so, from society itself. Once again, from this perspective, humans are treated as a mere "drag" on superior technology.

For many Futurists, human beings either function to aid the advance of technology (as inventors, scientists, or investors), or else they are a hindrance to this process (welfare recipients, state workers, and political malcontents). But if automation can produce durable goods more efficiently than the artisanal craftsperson, then perhaps the teacher, the nurse, the psychologist, or even the parent can be supplanted as well.

Robots and supercomputers can do it better: not only our jobs, but our very lives. A burgeoning genre in pop culture gives voice to this giddy "accelerationist" turn, as exemplified by the lyrics of the singer Grimes.

---

12. One should note, however, that not all self-described Futurists or accelerationist thinkers affirm capitalism, as Chapter 2 of this book will consider. A subset of this tendency identifies as anti-capitalist, if not socialist or communist.
13. The political viability of UBI came to prominence in the 2020 US presidential election, especially with the candidacy of Andrew Yang. Meera Jagannathan, "As Andrew Yang Drops out, Here's What Other 2020 Democrats Say about Universal Basic Income," *MarketWatch*, February 13, 2020, https://www.marketwatch.com/story/heres-what-2020-democratic-candidates-for-president-have-said-about-universal-basic-income-2019-07-09.

People like to say that we're insane
But A.I. will reward us when it reigns
Pledge allegiance to the world's most powerful computer
Simulation is the future[14]

On the surface, the eco-pessimist and the Futurist are indeed polar opposites. Their responses to modern crises appear to be diametrically opposed to one another. On the one hand, the eco-pessimist emphasizes the harm that humans have done to the natural world. Ego is the problem—our instrumentalizing of nature for human ends. On the other hand, the technophile praises boldness and ingenuity but laments the pitiful realities of human frailty.

In truth, however, both camps are dedicated to a kind of anti-humanism. Humans are either maliciously destructive of their environment or a drag on otherwise unlimited innovation and progress. The concept "human" and its centrality to our value-schemes are what must be overcome. We see here an identity of opposites: The eco-pessimist and the Futurist, each in their own way, desires a faithful "leap beyond" the human being. Neither camp is able to diagnose the present crisis as a product of the mundane realities of capitalist society. As such, they seek to solve our all-too human problems through a near-mystical self-transcendence.

Prometheus and Gaia: Two Conceptual Personae

Historically, such romanticism has been symbolized by two divine figures borrowed from ancient Greek myth: Prometheus and Gaia. Prometheus is famously that titan who stole fire from the Gods and gave it to humanity; Gaia is the primordial deity of Nature herself. These mythical figures, when deployed politically, are meant to express mutually exclusive attitudes toward the world and humanity's place in it. They are conceptual personae, namely, philosophical personality types.

The Promethean puts a premium on the will to power. The world is defined by chaos and hostility. Even the word "define" goes too far, for definition implies something static, solid, and stable. There is only the constant struggle to live, to innovate, and to conquer upon this hostile terrain—to win the prize of divine fire. Crises in the economy, like crises in ecology, are par for the course. These are not anomalies of a well-ordered Nature, but so many expressions of the underlying flux and chaos of things.

---

14. Grimes, *We Appreciate Power* (London: 4AD, 2018). It should be noted that the term "accelerationism" has a distinct, second meaning in the popular press. This involves extreme white supremacist ideology which seeks to "accelerate" a future race war.

What's more, such crises are opportunities for transformation. If the world is hostile to humanity, so much the worse for our parochial definitions of "humanity." We must aim, through our technologies and innovations, for the "post-human" future. All sentimental attachments to the past, to traditional value schemes, traditional forms of work, art, morality, and family should be sanguinely cast aside for the bright and open tomorrow that we can freely create for ourselves.

For the Gaian, these same crises elicit a diametrically opposed response. Climate change, rapid extinctions, and disruptions to traditional ways of life are distortions of a harmonious Whole. It is through our arrogant domination of Nature that we sow the seeds of our own destruction. Through the use of instrumental reason, we presume to know Nature, and so convert the world into a collection of "resources." But intensified exploitation, aided by new technologies, only serves to deplete the vital bases of human life. Ignoring that humans are but one species among many, we are rapidly making the planet inhospitable to our continued existence. Hubris will be our downfall. But perhaps our status as apex exterminators warrants such a fate.

The Promethean compels us to push past all limits; the Gaian counsels respect for natural boundaries and roles. The Promethean nominalizes nature and drains it of all inherent meaning. Nature is a nonsubstantial limit, there to be broken by the strong will of the creator, the inventor, and the industrialist. The Gaian, instead, theologizes Nature. We are not meant to understand Gaia, but to bow down and worship the Earth—to approach Nature in a spirit of reverence, not instrumental reason. But for all these stark differences, the Gaian and the Promethean are united in their misanthropy. Humanity (or humanism, or human reason) is the problem. For the Gaian, human exceptionalism breeds a self-destructive drive toward domination and violence; for the Promethean, flesh-and-blood "humanity" is an arbitrary limit on the unlimited powers of technology and invention.

These are highly speculative images drawing on ancient Greek lore. But, at the same time, they are fashionable responses to the politics of today. Terrified by the staggering problems we face, human nature is scapegoated and becomes the enemy. This is, in Marxist terms, the very definition of ideology, namely, the refusal to confront a problem directly.[15]

Of course, the images of Prometheus and Gaia could be usefully repurposed in nonideological ways. Prometheus can well represent human reason, ingenuity, and the capacity to improve one's state. Gaia could

---

15. For an in-depth overview of Marx and Engels's critique of ideology, see Jan Rehmann, *Theories of Ideology: The Powers of Alienation and Subjection*, Historical Materialism Series 54 (Chicago: Haymarket Books, 2014), 21–58.

represent a recognition of Nature as the basis for human life, and humans as but one modification of this intelligible whole. None of this would involve turning away from either reason or an empirical investigation of the world or class society. In the best Marxian tradition, we might recognize that "naturalism [...] equals humanism, and [...] fully developed humanism equals naturalism."[16]

Yet this is not how the personae of Prometheus and Gaia have been typically deployed, even in the present day. Here, an analytic distinction between the "mythic" and the "mystic" is useful. The mystic takes legends too seriously, confusing their fantastic appearances for the mundane realities they are meant to represent. In the case of Prometheus and Gaia, the supernatural elements of each persona are hypostatized and taken to be somehow real in themselves. We retain some residue of magical thinking, even as we know that these figures don't really exist. Promethean lore emphasizes a supernatural ability to overcome all material obstacles, and Gaia invokes the inscrutable, immaterial depths of existence. Even if the proper names of Greek scripture are discarded, these supernatural elements still have a hold on people's imaginations. We want to believe in a world of supernatural miracles beyond ordinary, natural laws. But giving independent life to the supernatural is a mistake. Such a flight into the fantastical only obscures the concrete challenges humanity faces.

The antidote to this "mystic" approach is a genuinely "mythic" sensibility. The mythic reading has the distinct advantage of actually understanding the function of ancient stories. Namely, myths, legends, fables, and holy Scriptures are nothing but tools for human use. They are not wellsprings of wisdom and truth but instead are reservoirs of vocabulary and images. These images are deployed by human beings as a means of communicating their fundamental worldviews. As the young Marx puts it, "it is not religion which creates man but man who creates religion."[17]

This recognition of how myth functions can be shocking. It is particularly discomfiting because it places a great burden of responsibility squarely on our shoulders. Once we realize that myths are not the source of truth but instead a means of poetically communicating our ideas, then we become answerable to those ideas we wish to express. No longer can we hide behind the authority of old stories as an excuse for being uncritical about our own truth-claims. The mythic attitude, then, emphasizes reason over belief and obedience. For

---

16. Karl Marx, *Economic and Philosophic Manuscripts of 1844* (Mineola, NY: Dover, 2012), 102.
17. Karl Marx, "Preface to a Contribution to the Critique of Political Economy," in *The Marx-Engels Reader*, ed. Robert C. Tucker, Second Edition (New York: W. W. Norton, 1978), 20.

it doesn't matter how eloquent one's speech is if the ideas it expresses are worthless.

The sine qua non of this position is the dictum that "there is no truth in religion." This is not a statement of simple atheism, but again, a recognition of what religion is—*an instrument or means for* conveying truths. Myths illustrate what we already claim to know but are never a means for "uncovering" new knowledge. This not only distinguishes the mythic thinker from fundamentalist religion, but also from those modern and postmodern substitutions for orthodox belief. The Jungian mystic, who speaks of timeless spiritual archetypes and the so-called collective unconscious, is no less guilty than the orthodox believer in this regard; each seeks ultimate truth beyond human reason within some realm of ineffable authority. The properly mythic thinker does not seek recourse in such collective unconsciousness, but instead, works toward collective consciousness, that is, the self-awareness and self-mastery that comes with seeing things clearly.

Therefore, the order of operations is of supreme importance. We must first get our philosophy right before we will know how to reconstruct our mythologies; we need to have the right ideas first before we can think of how best to express them. If this can be accomplished, then Prometheus and Gaia can function to illustrate the human and natural worlds as they really are. In other words, they can help constitute a mythology of Reason.[18]

Some might ask: Why bother to mythologize reason as all? Why cannot reason speak for itself? To this, we respond that rational discourse is indeed possible, even indispensable, without the pyrotechnics of mythic imagery. It is very important to be able to express one's fundamental worldview in clear and concise language, demonstrating claims using rational proofs.

At the same time, we are finite, bodily creatures and, as such, are subject to the whole range of human emotion. We operate according to images and sensations just as often (perhaps more often) than through pure, deductive reasoning. We therefore have a choice: We may give ourselves over to emotion and sensation, letting reason become a servant to our unaccountable passions. Alternatively, we could try to constantly repress our emotions, and pretend that they do not exist. But this repression, like all such attempts, will only serve to grant our passions greater strength and autonomy, and eventually, a return of their power over us. The third, and clearly preferable, option is to recognize that we are subject to the emotions and imagination, and to cultivate these for rational ends. We can subordinate our affects to the intellect, much in the

---

18. G. W. F. Hegel, "The Earliest System-Program of German Idealism," in *Miscellaneous Writings of G.W.F. Hegel*, ed. Jon Stewart (Evanston, IL: Northwestern University Press, 2002), 110–12.

way Socrates counsels in Book IX of the *Republic*: Our affects are like a many-headed beast, always changing, with some of the heads tame and social, and others antisocial.[19] It is up to us not to kill this beast (an impossibility), but to properly cultivate it and subordinate it to the command of Reason. Mythic images are but a product of our imagination. We can choose to use myth, rather than letting it use us.

This understanding of myth is anticipated in three great rationalist thinkers: Plato, Freud, and Brecht. In Plato's *Republic*, we see the call to replace bickering Homeric gods with more virtuous and rational exemplars.[20] No tears are shed for openly contravening the ancient stories; a bad tool is a bad tool, and needs replacing with a better one. Likewise, Freud sees therapeutic value precisely in uncovering those latent desires, traumas, and myths which influence our actions; the goal is to put these, as far as possible, under our own rational control rather than submit to their power over us.[21] For his part, Bertolt Brecht's concept of theater subordinates art to truth. Entertainment's function is to display the fundamental truths about society; materialist dialectics replaces the aristocratic idea of "art for its own sake."[22]

In their earliest manifestations, Prometheus and Gaia were not mythic but really mystic responses to contradictions in ancient class society. They did not communicate the actual conditions of exploitation, but merely expressed in supernatural garb the anguish which resulted from worldly oppression. (In other words, they obscured material realities rather than uncovering them.) While the mode of production in antiquity is a far cry from modern capitalism, the "status quo" of stratified class society is still with us today (albeit in different forms). It is for this reason that the uncritical use of Prometheus

---

19. Plato, *Complete Works*, ed. John M. Cooper (Indianapolis, IN: Hackett, 1997), Republic 588c.
20. Plato, *Complete Works*, Republic 376e–83c.
21. Hence, Freud was interested in the literary, that is mythic, personalities of Oedipus and Hamlet not because they were historical realities or because they were falsehoods in need of debunking; instead, the stories about these figures have instrumental value. In examining these character's psychological complexes, perhaps, we can become more aware and so cope with these same dynamics within ourselves. For an excellent discussion of this, see Peter Gay, *Freud: A Life for Our Time* (New York: W. W. Norton, 2006), 318–20.
22. Brecht often expressed a behaviorist understanding of art, and specifically theater. The point of theater was not merely to be beautiful or showcase certain emotions for their own sake, but rather to use affects in order to convey a point or elicit a certain attitude in the audience. See Bertolt Brecht, "The Threepenny Lawsuit," in *Brecht on Film and Radio*, ed. and trans. Marc Silberman (London: Bloomsbury, 2000), 147–202; Bertolt Brecht, *Brecht on Theatre: The Development of an Aesthetic*, ed. John Willett (New York: Hill and Wang, 1992), 150.

and Gaia persists. They are symptoms of a restlessness toward an oppressive social order. But rather than uncovering this order, they offer only an abstract and imagined form of revenge. Unable to conceive of a better organization for human society, an *inhuman* form of retaliation is imagined—either of autonomous technology (Prometheus) or revanchist Nature (Gaia). To understand Prometheus and Gaia today, one must explore the earliest ideological responses to class society.

Myth-Making and Class Society

In the beginning there was Hesiod. Or, at least, the Greek poet Hesiod gives us the earliest recorded myth involving both Prometheus and Gaia.[23] Since the First Agricultural Revolution, and likely before that, there were rulers and ruled, and there were stratified class societies. It is with Hesiod's poem, the *Theogony*, that we have one of the first religious justifications of established political order. In fact, while the poem begins with primordial concepts such as Chaos, Night, Sky, and Earth, it ends with a meticulous list of the supposed descendants of the gods which constituted the Greek aristocracy of Hesiod's time. In this way, the *Theogony* is a paradigm of what Marx calls ideological "superstructure," that is, material class relations reflected in subjective form.

Hesiod himself is a figure shrouded in mystery. Scholars debate whether he existed at all, or if he represents a collection of authors.[24] However, the way in which "Hesiod" presents himself is extremely revealing as to the ideological attitude behind his narratives (both the *Theogony* itself, as well as *Works and Days*). He is something of a rustic, living in the countryside, avoiding sea voyages, and issuing from a poverty-stricken family. One of the better-known lines from *Works and Days* is Hesiod's dismal portrayal of his hometown, Ascra: "bad in winter, godawful in summer, nice never."[25]

On the other hand, Hesiod has made good despite his poor circumstances. The reader learns this not through the author's direct bragging, but by the frequent excoriations directed at his good-for-nothing, profligate "jerk" of a brother.[26] *Works and Days* is full of practical advice and wise-sounding maxims,

---

23. Hesiod, *Works & Days. Theogony*, trans. Stanley Lombardo (Indianapolis, IN: Hackett, 1993). It should be noted that the figure of Gaia is indirectly referenced in Homer's *Iliad*, but is there not part of a coherent creation myth.
24. Hesiod, *Works & Days. Theogony*, 19. NB: References to *Works & Days. Theogony* (excluding the Introduction and Translator's Preface) are to the line number rather than page number. This will be indicated by listing the title of the relevant work followed by the line number.
25. Hesiod, *Works & Days. Theogony*, Works & Days 708–10.
26. Hesiod, *Works & Days. Theogony*, 2.

praises thrift, industry, honest-dealing, and clean living. For a poem originating from the early Iron Age, it is surprisingly bourgeois and middle class in its sensibility. Despite predating generalized commodity production, its value-schema anticipates something of the modern Protestant work ethic. Truly, Hesiod is the Horatio Alger of the ancient Mediterranean.

And much like Alger, Hesiod's denigration of the idle poor is supplemented by a "healthy" respect for authority and the well-to-do. There is the presumption that, at least in most cases, the rich owe their wealth to virtue, personal merit, and hard work—or at least rightful inheritance. Unlike Alger, however, Hesiod shows a greater propensity to criticize those powerful individuals who abuse their authority and station. These he describes in poetic terms as the "bribe-eating lords."[27] But for all that, Hesiod is quite enthusiastic to sing the praises of private accumulation and even economic competition. He even advances an early, rudimentary version of the invisible hand argument, whereby a certain kind of "strife" and egoism among individuals can produce a common good.

> When a person's lazing about and sees his neighbor
> Getting rich, because he hurries to plow and plant
> And put his homestead in order, he tends to compete
> With that neighbor in a race to get rich.
>
> Strife like this does people good.
>
> So potter feuds with potter
> And carpenter with carpenter,
> Beggar of beggar
> And poet of poet.

Such "feuding" is supposedly productive of greater skill and industriousness overall. Hesiod uses the Greek creation myth as a means to defend his fundamental worldview of authority, decency, industry, and, especially, law and order. For Hesiod, civilization is an evolution from anarchy, through a number of mediations, ending ultimately in legitimate structures of authority.

Thus, the first primordial entity Hesiod introduces in the *Theogony* is Chaos itself; and from Chaos emerges the Earth, personified as the goddess Gaia. In contrast to Chaos, Gaia represents a more determinate nature. She sets forth, from her own womb, the basic contours of physical reality. She gives birth to Uranus (Sky) as her equal and consort, as well as the Mountains and the Sea.[28]

---

27. Hesiod, *Works & Days. Theogony*, Works & Days 55.
28. Hesiod, *Works & Days. Theogony*, Theogony 126–33.

So much for the contours of the universe. What follows in the *Theogony* are three major phases in the cosmic balance of power. The initial equality between Gaia and Uranus does not last. Representing the sky, Uranus takes on the role of primordial kingship. He is the aloof sovereign who is ruling creation from above. But Uranus's rule is despotic; he will not admit of any differences, novelties, or variations in his domain. He hates the sundry offspring of Gaia and himself (i.e., the one-eyed Cyclopes and the hundred-handed Hecatoncheires). He cruelly shoves these monstrous children back into mother Gaia, causing her to groan.

> These monsters exuded irresistible strength.
> They were Gaia's most dreaded offspring,
> And from their start their father feared and loathed them.
> Ouranos [Uranus] used to stuff all of his children
> Back into a hollow of Earth [Gaia] soon as they were born,
> Keeping them from the light, an awful thing to do,
> But Heaven [Uranus] did it, and was very pleased with himself.[29]

Here we have the unyielding power of the father as against women and children. More abstractly, we have the domination of sameness and uniformity as against unruly difference (the "fear and loathing" of the other). Uranus's kingship is based purely upon the direct, brutal assertion of power. His power, however, is brittle, for it lacks law and legitimacy.

Mother Gaia thus conspires with her child Cronus against their overbearing father. She equips him with a serrated sickle. Cronus, using stealth and trickery, ambushes his father and castrates him. His genitals drop to the sea, and from the resulting waves emerge Aphrodite. This goddess of love and beauty symbolizes a reemergence of the sacred feminine.

In this second phase of power, Cronus ends the suffocating rule of uniform Being and ushers in the rule of Time. He is often depicted with his scythe, which not only hearkens to this episode of filial rebellion, but also represents agriculture and the seasonal harvest. Reaping, destruction, and the passage of time are all correlated in this mythic figure, as revealed during the consolidation of his kingship.[30]

---

29. Hesiod, *Works & Days. Theogony*, Theogony 153–59.
30. Hesiod does not explicitly equate Cronus with Chronos (Time). Nonetheless, the reception of this myth through Greek and Roman antiquity has perennially upheld their identity. This can be seen, for instance, in Plato's *Cratylus* as well as in Cicero's *On the Nature of the Gods*. See Plato, *Complete Works*, Cratylus 402b; Cicero, *De Natura Deorum and Academica*, ed. E. H. Warmington, trans. H. Rackham, vol. XIX (Cambridge, MA: Harvard University Press, 1967), 185.

At some point, Cronus learns of a prophecy that, just as he overthrew Uranus, one of his children would one day depose him in turn. He thus resolves to eat all of his offspring rather than letting them mature and take power. This resembles the actions of his father, with one crucial difference: Whereas Uranus could not permit the existence of any particulars apart from himself (shoving his children back into Gaia), Cronus actively devours them. In the former case, Being (Uranus) is the pure positivity which does not allow of any difference whatsoever; in the latter case, Time or annihilation (Cronus) is the pure negativity which consumes all differences within itself. With Cronus, the other is not merely excluded; it is relentlessly, mercilessly overcome.

In this way, both kingships are brutal and one-sided. As Hegel put it, "pure light and pure darkness are two voids that amount to the same thing."[31] Nonetheless, the rule of Cronus is something of an advance over that of his father, Uranus. (Or, at least, this is its mythic function.) Time/negation/motion are more determinate and productive concepts than bland, uniform Being. Cronus thus stands as a middle term between monotony and earthly politics. One cannot go from Uranus to the Greek city states without passing through Cronus (time, change, seasons, agriculture, and violence) first.[32]

In the *Theogony*, the negation is then negated, and the trickster gets tricked. Cronus's wife Rhea is upset that all of her children are being devoured by her husband. Once her youngest child Zeus is born, she entrusts him to his grandmother Gaia. Gaia hides the young child in a cave while Rhea deceives Cronus; he is given a stone covered in swaddling clothes as a substitute Zeus, which he then swallows—thus sparing the infant god.

> Down into his belly, the poor fool! He had no idea
> That a stone had been substituted for his son, who,
> Unscathed and content as a babe, would soon wrest
> His honors from him by main force and rule the Immortals.
> It wasn't long before the young lord [Zeus] was flexing his glorious
>   muscles.[33]

---

31. G. W. F. Hegel, *The Science of Logic*, ed. and trans. George di Giovanni (Cambridge: Cambridge University Press, 2010), sec. 152.
32. This is analogous to the blade-wielding figure of Kali within the Hindu pantheon. Kali embodies cyclical change or destruction, but also those changes which are productive of new life (as within patterns of agriculture). Kali's name is derived from the Sanskrit word for time itself.
33. Hesiod, *Works & Days. Theogony*, Theogony 491–95.

Thus begins the third shift in cosmic rule. Zeus grows into a physically powerful deity, the leader of the Olympian gods. The mythic pattern holds, and Gaia once again sides with the upstart as against the regime. She helps Zeus defeat Cronus, and in turn, Cronus disgorges all of Zeus's siblings. Then Gaia counsels Zeus to liberate the Cyclopes and the hundred-handed monsters (Hecatoncheires) in order to aid in the fight against Cronus. This war between the Olympian gods and the Titans (the "Titanomachy") represents the founding of a positive, established order. The rule of Olympus under Zeus mirrors the aristocratic rule of the Greek *poleis* (city-states).

The mythic repetition then continues with a still newer conflict between Zeus and Gaia. As with the rulers before him, Zeus threatens to suppress all contenders for his throne. According to the ancient commentator Apollodorus, Gaia is angry with Zeus for the destruction of her children, the giants.[34] She births the sea monster Typhon as a means of revenge. The rule of Olympian order is challenged by the paradigmatic image of chaotic nature—the water beast. Cognates of this in world mythology abound, from the Babylonian Tiamat (described as a saltwater she-dragon) to the female Leviathan in the Book of Enoch, as well as similar primordial water beasts in Job 41, Isaiah 51:9–11, and Psalm 104:26.[35]

But here the repetition is interrupted. Instead of the challenger (Typhon) overturning the existing order (as with the previous defeats of Uranus and Cronus), Zeus roundly vanquishes Typhon; Nature is definitively subdued by the rightful order of Olympus. This parallels the episode in Psalm 74:12–21 wherein Yahweh defeats the sea monsters and establishes the durable order of heaven for the benefits of his chosen people.

> But God is my King from long ago;
> he brings salvation on the earth.
> It was you who split open the sea by your power;
> you broke the heads of the monster in the waters.
> It was you who crushed the heads of Leviathan
> and gave it as food to the creatures of the desert.

---

34. Apollodorus, *The Library*, trans. James George Frazer (Cambridge, MA: Harvard University Press, 1921), sec. 1.6.3. As Plato suggests, the giants represent the material, bodily, and natural, as opposed to the Gods which represent intelligible Forms. Plato, *Complete Works*, Sophist 246a.
35. R. H. Charles, *The Book Enoch* (London: London Society for Promoting Christian Knowledge, 1917), 60:7; Stephanie Dalley, trans., *Myths from Mesopotamia: Creation, The Flood, Gilgamesh, and Others*, Revised Edition, Oxford World's Classics (Oxford: Oxford University Press, 2000), 228–77.

It was you who opened up springs and streams;
you dried up the ever-flowing rivers [...]
Have regard for your covenant,
because haunts of violence fill the dark places of the land.
Do not let the oppressed retreat in disgrace;
may the poor and needy praise your name.[36]

After Zeus's ultimate victory, the various functions of Nature are taken over and divided among the Olympians. Zeus takes the thunderbolt from Uranus, Poseidon takes the sea along with the trident, while Hades reigns over the underworld (death). In nearly all cases, the fundamental forces of nature are expropriated by anthropomorphic figures. They are put to human-like ends rather than left to the various Titans, giants, monsters, and primordial entities associated with Gaia before the rule of Zeus.

With the defeat of Typhon, her last-born child, Gaia groans once more:

When Zeus' temper had peaked he seized his weapons,
Searing bolts of thunder and lightning,
And as he leaped from Olympos, struck. He burned
All the eerie heads of the frightful monster.
And when he had beaten it down he whipped it until
It reeled off maimed, and vast Earth groaned.[37]

What follows is apocalyptic-like language, describing the far-reaching destruction of the Earth (Gaia) as it was previously known. This parallels certain aspects of the Book of Revelation. Here, the triumph of the forces of good over the devil-serpent (also thrown into an abyss) accompany a remaking of the natural landscape.[38] In Hesiod's telling, this refashioning of the world is closely tied to human *techne* and industry. The hot fury of Zeus symbolizes the Greek Iron Age, with its bellows and furnaces.

And a firestorm from the thunderstricken lord [Zeus]
Spread through the dark rugged glens of the mountain,
And a blast of hot vapor melted the earth like tin
When smiths use bellows to heat it in crucibles.
Or like Iron, the hardest substance there is,
When it is softened by fire in mountain glens

---

36. Ps. 74:12–21 NIV.
37. Hesiod, *Works & Days. Theogony*, Theogony 860–65.
38. Rev. 20:3 NIV.

And melts in bright earth under Hephaistos' hands.
So the earth melted in the incandescent flame.
And in anger Zeus hurled him [Typhon] into Tartaros' pit.[39]

With Gaia's groan of resignation, her role changes entirely. Thoroughly subdued, the humbled deity becomes something of a handmaiden and advocate for Zeus. No longer will she support younger challengers as against the reigning monarch. No longer will she use trickery to defend her vulnerable descendants against the same.

Now, fully domesticated, she supports Zeus's rule on Olympus as the leader of the Immortals. Gaia even reverses her role entirely, counseling Zeus to consume his pregnant wife, Metis, so that none of his talented children could one day overthrow him. The creation legend comes full circle: Zeus assumes the lordship over the sky, as his grandfather Uranus once did. But this time, the Earth will support no further rebellions, and his rule will have no end. This is no reconciliation of Order with Nature. It is rather the one-sided victory of determinate authority over the Earth.

That's not to say that this is the end of creation. New gods are yet to be born. Within Zeus, the pregnant Metis bears Athena. But, according to Hesiod, it is Zeus himself who bears his daughter from the top of his own head.[40] This is the triumph of the male principle over the feminine, and the rule of *logos* over the material and the bodily. After all, Athena's birth is not only unconventional because it was from a male, but also unusual in terms of the part of the body from which she sprang.[41]

Meanwhile, Athena's mother, Metis, plays a relatively passive role. It is unclear if she ever escapes Zeus or is simply consumed as raw material within his body. Metis was a pre-Olympian Titan and an Oceanid to boot. Zeus thus, again, overcomes the fluid, anarchic forces of Nature, cast as the primordial feminine. This account underscores the whole movement of Hesiod's *Theogony*: It is the passing of cosmic rule from Chaos, to Being, to Time, to Order.

---

39. Hesiod, *Works & Days. Theogony*, Theogony 866–74.
40. Hesiod, *Works & Days. Theogony*, Theogony 904–29.
41. In later versions, such as Pindar's *Olympian*, Zeus does bear Athena because of the pain he feels from her metal armor and helmet. As with Hesiod, Pindar casts the male principle as the active one. In Pindar's account, Zeus enlists the help of Hephaestus (the god of blacksmiths, craftsmen, of weapon-making, and metallurgy) to crack his head open with an axe in order to let Athena out. See William H. Race, *Pindar: Olympian Odes. Pythian Odes.*, Loeb Classical Library (Cambridge, MA: Harvard University Press, 1997), Olympian 7: 35–40.

This should not be understood as a timeless myth, but one entirely bound up with the practical concerns of its day. Writing at the cusp of the so-called dark ages and the first hints of the new city-states (*poleis*), Hesiod has a definite preoccupation with order and chaos. His work does double duty: On the one hand, it is an explicit apologia for worldly, political authority—connecting the durable rule of Zeus on Olympus with the temporal rule of Greek sovereigns and political dynasties.

But, on the other hand, this conscious apologia casts a dark shadow. This is the denigration of women who stand in for the forces of political and social regression, anarchy, and dissolution. It follows a well-trodden pattern common to earlier Babylonian lore; here, the female serpent Tiamat threatens existence through her chaotic restlessness. It takes Marduk, the king-god, and patron of the city of Babylon, to vanquish Tiamat and establish a universe habitable by human beings. Interestingly, he accomplishes this in a very similar way to Zeus's destruction of the serpent Typhon, namely smashing Tiamat's head with a mace. Women being associated with disorder, disobedience, and regression are readily seen in Abrahamic texts as well, notably the figures of Eve and Lilith.[42] In Hesiod, the only female characters who escape such vilification are those who are radically masculinized. These include the war-like and crafty Athena, and those who are sufficiently subdued and chastened, such as Gaia, following Zeus's consolidation of Power.

Nevertheless, Hesiod treats the male rivals to Zeus in a wholly different manner. Hesiod introduces the deity Prometheus with a flourish of complimentary adjectives: "complex, his mind a shimmer." This cunning trickster represents the challenge posed to Olympian authority. Immediately, Prometheus is described as being punished by Zeus for his excessive arrogance and individualism. Infamously, he is bound to a large rock, his liver being constantly devoured by a bird of prey.

> And he bound Prometheus with ineluctable fetters,
> Painful bonds, and drove a shaft through his middle,
> And set a long-winged eagle on him that kept gnawing
> His undying liver, but whatever the long-winged bird
> Ate the whole day through, would all grow back at night.[43]

---

42. Gen. 3:6 NIV; Howard Schwartz, ed., *Lilith's Cave: Jewish Tales of the Supernatural* (Oxford: Oxford University Press, 1988), 120–22. The figure of Lilith has since been reappropriated for feminist ends, for example the Jewish-feminist publication, *Lilith* magazine.
43. Hesiod, *Works & Days. Theogony*, Theogony 523–27.

But unlike the submission of Gaia, or the destruction of her serpentine child, or the consumption of Metis, Prometheus is eventually allowed to escape his punishment. This is for the benefit of yet another masculine character—Zeus's son Herakles. The latter performs the "heroic" rescue mission so that his reputation among mortals would be increased.

> Herakles, killed, [and] drove off the evil affliction
> From Iapetos' son [Prometheus] and freed him from his misery —
> Not without the will of Zeus, high lord of Olympos,
> So that the glory of Theban-born Herakles
> Might be greater than before on the plentiful earth.
> He valued that and honored his celebrated son.[44]

It seems that Zeus not only allowed for this stay of punishment for the benefit of his own son, but also because he admired the haughty and clever Prometheus who "matched wits" with him.[45]

There is clearly a gendered double-standard at work here. The anarchic forces of femininity are to be permanently quashed. This is a prerequisite for the stable rule of Olympus. On the other hand, the challenge of crafty, arrogant, and prideful males is respected, if not always tolerated. These are rivals to be admired, rather than feared or totally obliterated. Zeus plays with Prometheus, sometimes punishing him, at other times letting him go, always keeping up this masculinist state of "play."

Representative of this dynamic was the so-called Trick at Mekone episode. At a time when the gods and men were setting down the conventions for sacrificial rites, Prometheus intervened and slaughtered a large ox. He divided the animal into two portions, one to be chosen by Zeus (representing the gods' share of sacrifices), the other to be left to mortal men. The one portion consisted of choice pieces of meat but concealed within itself the unappetizing stomach of the ox. The second portion consisted of mere bones, but was concealed within the alluring, glistening fat of the animal.

Zeus, described by Hesiod as supremely wise, immediately saw through such trickery. Yet instead of calling Prometheus out, he plays along with the ruse. Zeus chooses the inferior portion and then uses this as a pretext to undo Prometheus's advocacy for human beings. While humans will, from this point on, only have to sacrifice the poorer parts of the animals to the gods, Zeus also removes the power of "weariless fire" from mortals.

---

44. Hesiod, *Works & Days. Theogony*, Theogony 523–27.
45. Hesiod, *Works & Days. Theogony*, Theogony 536.

The rivalry then continues, and Prometheus famously resolves to steal fire back from Zeus in turn.

> But that fine son of Iapetos [Prometheus] outwitted him [Zeus]
> And stole the far-seen gleam of weariless fire
> In a hollow fennel stalk, and so bit deeply the heart
> Of Zeus, the high lord of thunder, who was angry
> When he saw the distant gleam of fire among men,
> And straight off he gave them trouble to pay for the fire.[46]

This "trouble" is woman. Here, the misogynistic tropes of Hesiod's myth come to a crescendo. Enter the figure of Pandora.[47] She is an instrument of Zeus to dissipate man's energies. "Weariless fire" is the means by which mortal men accomplish their work and sustain themselves. With fire, we cook food and work metal, and generally distinguish ourselves from other animals. But Zeus sets out to confound our prosperity. Once again teaming up with the forge-god Hephaestus, he orders the creations of an artificial woman, Pandora, to afflict human beings.

Outwardly, Pandora is seductive in her apparent innocence. She is covered in a maiden's veil and fashioned with the face of an immortal goddess.[48] Pandora is then granted a figure "like a beautiful, desirable virgin's." Athena teaches her the feminine arts of embroidery and weaving, and Aphrodite gives her the power to induce "painful desire" and "knee-weakening anguish" in men.[49]

But appearances are deceiving. While Pandora looks like a "shy virgin," adorned in fine jewelry and springtime flowers in her hair, she is given, according to Hesiod, "a bitchy mind and a cheating heart."[50] She is explicitly described as the "irresistible bait" that Zeus has laid out to trap mankind.[51]

Pandora is first offered to the dimwitted Epimetheus ("Hindsight," the brother of Prometheus). True to his name, he accepts her and only later considers the dire consequences for humanity. In perhaps the most famous episode from the *Theogony*, Pandora lets loose all of the terrors and misfortunes which plague the mortal world to this day. It is because of this that life is so

---

46. Hesiod, *Works & Days. Theogony*, Theogony 569–72.
47. She is referred to as "The Maiden" in *Theogony* and called by her proper name in *Works and Days*.
48. Hesiod, *Works & Days. Theogony*, Theogony 577.
49. Hesiod, *Works & Days. Theogony*, Works & Days 80–90.
50. Hesiod, *Works & Days. Theogony*, Works & Days 87.
51. Hesiod, *Works & Days. Theogony*, Works & Days 103.

hard and perilous, and that we must constantly labor just to sustain ourselves. There are obvious parallels between this element of the poem to Genesis 3:23. There, too, a woman is the instrument of man's downfall and his requirement to labor. "So the Lord God banished him [Adam] from the Garden of Eden to work the ground from which he had been taken."

Crucially, Pandora is not a singular figure, like Typhon or some other threatening monster; rather, she prefigures all mortal women: "from her is the race of female women, the deadly race and population of women, a great infestation among mortal men."[52] Her plagues are not only singular and supernatural, but rather permeate male/female relations in general. Women are described by Hesiod as drags on male productivity. They consume, but do not work; they lack all sense of thrift, being "at home with wealth but not poverty." By way of analogy, he compares them to sterile (i.e., "unproductive") drones in a beehive. The male worker bees toil from dawn till dusk in order to feed themselves as well as the useless drones. The latter are always "stuffing their stomachs with the work of others."[53]

In the *Theogony*, Pandora and the resulting race of women are used by Olympus in a very peculiar way which reveals the underlying worldview of Hesiod himself. Women do not utterly destroy men, but rather, they keep the race of man in check relative to their superiors, the gods. They are the instruments of enforced humility. "He [Hephaestus] made this lovely evil to balance the good."[54]

Indeed, the best we can hope for in this world is a life of toil, compromise, strife, and struggle—or in other words, marriage. For Hesiod, a good marriage (ostensibly between a man and a woman) is something of a midway point between heaven and hell. For, "whoever marries as fated, and gets a good wife, compatible, has a life that is balanced between evil and good, a constant struggle."[55] True, women are unproductive consumers, but at least they produce children. A man without children (or, more specifically, sons) is one who is doomed to a horrendous, unsupported old age, and one whose estate is doomed to be divided up among distant relatives or strangers.[56]

---

52. Hesiod, *Works & Days. Theogony*, Theogony 595–96.
53. Hesiod, *Works & Days. Theogony*, Theogony 603. Of course, we know today that idle drones within honeybee hives are actually male, and their biological function is to mate with the unfertilized queen. Hesiod was writing in a largely prescientific context and, in any case, his poem is more of a misogynistic polemic than a zoological textbook.
54. Hesiod, *Works & Days. Theogony*, Theogony 588.
55. Hesiod, *Works & Days. Theogony*, Theogony 611–14.
56. Hesiod, *Works & Days. Theogony*, Works & Days 772–77.

Between a rock and a hard place, men are forced to acquiesce to marriage. But Hesiod's writings make it clear: Marriage is not ordained by the gods as a sublime good, but rather as a concession to the realities of material scarcity. There is a constellation here, the elements of which are labor, order, and obedience. The male head of the household is tethered to a woman for his entire adult life on pain of an unsupported old age. Women are made to submit to their husbands for their basic sustenance. And the subordination of both wives and children to male householders mirrors the subordination of men to their social betters. What keeps this discomfiting picture together, according to Hesiod, is basic material necessity. Why else would one submit to a lifetime of mutual obligation and servitude?

But not everyone will be satisfied with the rule of Olympus, and more importantly, what it represents. Resentment against the constituted order will elicit its own set of myths. A stratified class society, such as Iron Age Greece, will be based on the direct, often brutal exploitation of labor. Both economic and political power will be concentrated in a few hands, while the masses will be in a permanent state of disenfranchisement.[57] There will be some who wish to escape both their condition of labor, and the domestic bonds which enforce this condition. Such individuals will seek their personal liberty, both from lord and from family.

This individualistic rebellion, against very real forms of oppression, is embodied in the Promethean spirit. It is the desire to supersede material and social bonds with an unlimited sort of freedom. Others, inspired by Gaia, will plot their revenge against Olympus in another way entirely: They want to humble authority, law, and civilization before a titanic Nature.

This book is a critical investigation into these two ideological tendencies. It will be demonstrated that each is a form of anti-humanism. They cast the human being as either a fetter on unlimited freedom and production, or else as a violent burden upon the Earth. Chapter 1 will trace the legacies of Prometheus and Gaia up through modernity. Chapters 2 and 3 will analyze the contemporary philosophies of Prometheus and Gaia, understood today as Accelerationism and Eco-Pessimism, respectively. The surprising identity between these two positions will be explored in Chapter 4. This will be followed by a suggestion for rehabilitating the figures of Prometheus and Gaia, inspired by the works of Aeschylus, Percy Shelley, and Karl Marx. Such a new mythos will affirm humanism, naturalism, and democracy as mutually compatible ideals.

---

57. For an excellent discussion on stratified class society in Greek antiquity, including a discussion of Hesiod in particular, see Geoffrey Ernest Maurice de Ste. Croix, *The Class Struggle in the Ancient Greek World: From the Archaic Age to the Arab Conquests* (London: Duckworth, 1981), 278.

Chapter 1

PROMETHEUS AND GAIA

Myths are always a reflection of personality. The myths we create, accept, reject, and tell others perhaps tells us most about ourselves. But mythic tropes do not point to one, invariable, mystical Truth of the human condition. Instead, they are born out of our human need to narrate common experiences. These experiences may be more or less universal, ranging from the biological processes of birth, adolescence, reproduction, and death, to the highly situated conditions of life within this or that class society. The myths we create and share are always value-laden. For they reflect, at least implicitly, the normative positions of the storyteller. Myths, therefore, are narratives as well as diagnoses. This is the case with the mythic figures of Prometheus and Gaia.

These images began their recorded existence in Iron Age Greece as part of a *Chaoskampf* story—a narrative about the creation of order out of chaos. Depending on the storyteller and context, Prometheus and Gaia have evolved, been modified, and repurposed throughout the ages. Still, a degree of their original, Greek significance has remained, attesting to their powerful hold over the human imagination. Prometheus has long stood for mastery, heroism, innovation, technology, and progress, as well as defiance against sclerotic order. Gaia, for her part, perennially evokes the notions of balance, vitality, care, and respect for a primordial Being which precedes civilization.

In this chapter, we shall explore how these figures have appeared in literature and popular culture. We will also demonstrate a tendency for these titanic personae to turn anti-humanist, especially when human storytellers are subjected to the unique pressures and traumas of modern capitalism. A leap to the "titanic" is often an expedient for the hopeless and scared. In any case, social being—how we materially produce and reproduce society—determines mass consciousness, and consciousness includes myth. But which myths we affirm can tell us a lot about how we hope to cure society's ills.

Enter Prometheus

The Promethean persona is born of an individualistic rebellion. It is resentful, not only of the boss, but also of the family. It reacts, not only against the established order, but against the vulnerable as well. To the Promethean imagination, women and children represent need, care, dependency, and above all, unproductivity. In the works of Hesiod, the Olympian order uses the family as a means to control mankind. Natural needs and natural limits protect the ruling elites as against those strong souls who would become free. Matter, the difficulty of labor, and the fact that there are so many mouths to feed all stand in the way of a boundless liberty. "Prometheanism" means stealing from the gods, not only ordinary fire, but the secrets of supermundane productivity. It means, ultimately, frictionless production without material or labor inputs. It is independence from all social bonds and natural limits.

> You know, the gods never have let on
> How humans might make a living. Else,
> You might get enough done in one day
> To keep you fixed for a year without working.[1]

In both *Works and Days* and the *Theogony*, Prometheus is described as intelligent, "his mind a shimmer," but also as devious and "shifty."[2] While not quite a villain, he is at most an incorrigible antihero. Prometheus does not submit to the rightful rule of Zeus, whom Hesiod never fails to praise as the wisest and most just of the gods.

The Satanic Prometheus

Prometheus is a Luciferian character—a titan who falls from grace, offering mankind both knowledge and a short-cut to worldly prosperity. And just as there have been perennial fascinations with Satan (especially among those dissatisfied with the established order), so too has there been a long-lasting and diverse tradition of Prometheanism. Both figures become the embodiments of human civilization and autonomy, fire and the fruit of knowledge being nearly interchangeable tokens of this promise.

Even the Russian anarchist Mikhail Bakunin, an avowed atheist, praises the figure of Satan as a liberating Promethean force. In his pamphlet *God and*

---

1. Hesiod, *Works & Days. Theogony*, Works & Days 58–61.
2. Hesiod, *Works & Days. Theogony*, Theogony 513; Works & Days 66.

the State (1882), he casts the Garden of Eden not as a paradise, but as a prison designed to enforce our prolonged dependency.

> But here steps in Satan, the eternal rebel, the first freethinker and the emancipator of worlds. He makes man ashamed of his bestial ignorance and obedience; he emancipates him, stamps upon his brow the seal of liberty and humanity, in urging him to disobey and eat of the fruit of knowledge.[3]

But Satan is no humanitarian—at least not in the Christian imagination. He tempts Eve with knowledge in order to spite God, not to relieve the estate of Man. This boon of knowledge, infamously, comes at the price of hard labor and the pains of childbirth, shame, and death.

The same goes for the account of Prometheus. His motives are suspect. Does Prometheus steal fire back from Zeus merely to aid in human flourishing? Or is the real aim of Prometheus merely to undermine the ruler of Olympus? It appears that one driver of his actions is a personal resentment against Zeus. Like his titan brother Atlas, Prometheus was punished for his cockiness and pride. Liberating humanity from servile obedience to the gods may just be an indirect way of reversing Zeus's domination.

But domination, by the gods, the titans, or among human beings, appears to be a durable state of affairs in Hesiod's worldview. In his "fable for the kings," he relates the hawk's speech to its nightingale prey:

> No sense in your crying. You're in the grip of real strength now,
> And you'll go where I take you, songbird or not.
> I'll make a meal of you if I want, or I might let you go.
> Only a fool struggles against his superiors.[4]

The presumption of predator and prey, master and slave, conqueror and conquered is the persistent default of this worldview. There is no sign that the titan Prometheus is an exception to this rule, namely, that he is a believer in the principle of equality for its own sake. This would, in any case, grate against his vaunted reputation as a clever and powerful trickster. In the end, Prometheus "the fire bearer" has much in common with Lucifer "the light bearer."[5]

---

3. Mikhail Bakunin, *God and the State* (New York: Dover, 1970), 10.
4. Hesiod, *Works & Days. Theogony*, Works & Days 240–43.
5. Raphael Jehudah Zwi Werblowsky, *Lucifer and Prometheus: A Study of Milton's Satan* (London: Routledge, 2001).

The Promethean worldview shows up repeatedly in various mythological contexts beyond the works of Hesiod. Genesis' account of the Tower of Babel is one such example.[6] Here, multitudes of people collaborate to build a great structure, and fearing for his supremacy, God confounds their common language, and so dooms the project. It is with the commentary of Flavius Josephus, however, that the truly Promethean elements of the Babel story come to the fore.

Josephus claims that it is Nimrod, a tyrant, who directs the building of this immense structure. Nimrod's motivation is to spite Yahweh by convincing people that they can accomplish this project "through his means" and "through their own courage" without the aid of God.[7] The tower, moreover, is meant to be so sturdy and tall that it could withstand another flood, and thus circumvent the terrible justice of the Lord. This classical interpretation is backed up by other first-century contemporaries, such as Philo, who see the description of Nimrod as "a giant before God" indicating his opposition to Yahweh.[8]

This interpretation finds continuation in modern times. For example, Dostoyevsky further criticizes the building of the tower for its Promethean aspirations. For him, modern socialism is prefigured by this arrogant act. Just as with Nimrod, socialism is inherently hostile to the authority of God, seeking to secure human happiness and esteem in this world by our own means alone.

> For socialism is not merely the labour question, it is before all things the atheistic question, the question of the form taken by atheism today, the question of the tower of Babel built without God, not to mount to Heaven from earth but to set up Heaven on earth.[9]

The Modern Prometheus: From Bacon to Shelley

With the Scientific Revolution, the figure of Prometheus obtains a more complicated hearing. One of the first modern writers on Prometheus was the English philosopher Francis Bacon. Bacon credits Prometheus as the maker and master of humanity, while also warning against his excesses. Prometheus's

---

6. Gen. 11:1-9 NIV.
7. Flavius Josephus, *Jewish Antiquities*, Wordsworth Classics of World Literature (Hertfordshire, UK: Wordsworth Editions, 2006), 16.
8. James L. Kugel, *Traditions of the Bible: A Guide to the Bible as It Was at the Start of the Common Era* (Cambridge, MA: Harvard University Press, 2009), 230.
9. Fyodor Dostoevsky, *The Brothers Karamazov*, trans. Constance Garnett (New York: Random House, 2012), 26.

interest in humanity is one of egotism; they are his pet project of which he is "extravagantly vain and proud."[10]

In *The Wisdom of the Ancients* (1609), Bacon emphasizes humanity's ingratitude toward Prometheus and his (albeit self-serving) gift of fire. He imagines humanity calling Prometheus to account before the judgment of Zeus. Human beings do not wish to be "liberated" by new technologies, but instead long for the old ways of doing things, even if this means continued domination by their social superiors. This is in line with Spinoza's lamentation in the *Theological-Political Treatise*, that an ignorant humanity will "fight for their servitude as if they were fighting for their deliverance."[11]

Prometheanism represents not only discovery, invention, and the removal of obstacles, but also an exhausted state of busyness. The "wise and fore-thoughtful class" of scientific men may relieve their estate for a time, but with this genius comes a definite neuroticism, a "torment and wear" and a preoccupation with "solicitude and inward fears."[12] The punishment of Prometheus (the eagle's consumption of his liver) thus symbolizes the torment of scientific man.

> For being bound to the column of Necessity, they are troubled with innumerable thoughts (which because of their flightiness are represented by the eagle), thoughts which prick and gnaw and corrode the liver: and if at intervals, as in the night, they obtain some little relaxation and quiet of mind, yet new fears and anxieties return presently with the morning.[13]

But speculation is not only worrisome; it can also be violent and brutal. The final crime of Prometheus, according to Bacon, is the attempted rape of Athena. As Athena represents divine wisdom, this crime is the ultimate act of hubris. For Bacon, we err when we try to "bring the divine wisdom itself under the dominion of [worldly] sense and reason." Mundane knowledge is not itself sinful, so long as we "distinguish between things divine and human, between the oracles of sense and faith."[14] In other words, we must be clear

---

10. James Spedding, Robert Leslie Ellis, and Douglas Denon Heath, eds., *The Works of Francis Bacon*, Volume VI (London: Longmans, 1890), 745.
11. Baruch Spinoza, *Theological-Political Treatise*, ed. Jonathan Israel, trans. Michael Silverthorne and Jonathan Israel, Cambridge Texts in the History of Political Thought (Cambridge: Cambridge University Press, 2007), Preface, par. 6.
12. Spedding, Ellis, and Heath, *The Works of Francis Bacon*, VI:751. "Fore-thought" is the literal translation of the name "Prometheus."
13. Spedding, Ellis, and Heath, *The Works of Francis Bacon*, VI:751–52.
14. Spedding, Ellis, and Heath, *The Works of Francis Bacon*, Volume VI:753.

as to the difference between ordinary, empirical insights which are proper to our limited intellects on the one hand, and the presumption of universal, total enlightenment on the other. An unlimited sort of reason, which ignores these boundaries of decency, is thus equated with rape.[15]

Prometheus (and the Promethean spirit) can be properly tamed; but this ability to moderate is, likewise, a gift from the gods. For "this virtue [of forbearance] was not natural to Prometheus, but adventitious [...] it is not a thing which any inborn and natural fortitude can attain to."[16] Therefore, Bacon's praise for humility is redoubled: Not only must we shed our hubris in order to live a decent and untroubled existence, but moreover, we cannot attain humility by our own efforts. We must beg for this gift from the gods themselves.

If this were all Bacon had to say about Prometheus, his views would be unremarkable—simply another addition to the "Babel" literature directed against the arrogance of humanity. But in *The Wisdom of the Ancients*, there is an intriguing shift which takes place. Yes, humanity complains to Zeus about Prometheus's theft of fire. But this is not a straightforward condemnation of knowledge as such.

In Bacon's hands, the indictment of Prometheus actually represents an impulse toward ever greater learning. Should humanity have accepted Prometheus's gift without complaint, this would amount to a childlike and uncritical reception of new technology. But such a pacified, juvenile acceptance would be one of self-satisfaction. This simple-minded arrogance would be akin to the greedy child grabbing gifts from under the Christmas Tree which they neither understand nor appreciate.

Bacon therefore praises the complaints of the crowd against Prometheus. These complaints show the crowd's modesty and hesitance in accepting new technological innovations. But rather than this amounting to a simple turning away from progress, such hesitation reveals mankind's critical, and therefore mature, reaction to innovation.

> They [...] who arraign and accuse nature and the arts, and abound with complainings, are not only more modest (if it be truly considered) in their sentiment, but are also stimulated perpetually to fresh industry and new

---

15. Bacon's association between unlimited knowledge and rape is revealing. It contrasts with popular feminist criticisms of Bacon himself as affirming an unlimited domination of nature. The truth, it seems, is that Bacon was quite hesitant to give knowledge free rein. For such feminist criticisms, see Sandra Harding, *The Science Question in Feminism* (Ithaca, NY: Cornell University Press, 1986), 113.
16. Spedding, Ellis, and Heath, *The Works of Francis Bacon*, VI:752.

discoveries [...] [T]he accusation of Prometheus, our maker and master though he be, yea sharp and vehement accusation, is a thing more sober and profitable than this overflow of congratulation and thanksgiving.[17]

For all this talk of modesty, Bacon still places mankind at the center of a providential universe. Without Man, he says, "the rest would seem to be all astray, without aim or purpose."[18] Despite his standing as a purely empirical philosopher, Bacon's teleological thinking is ubiquitous throughout his work. The sky exists so that Man can predict the weather; the wind so that he can fill his sails; flora and fauna so that he can eat, produce medicines, and other comforts, and so on.[19]

Bacon's reputation as a Promethean thinker is thus partly deserved. This is especially the case when it comes to his denigration of consumption and pleasure. The true purpose of our reason is to dominate nature, but not necessarily to enjoy it. Here, Bacon relates a story wherein the gift of everlasting life is placed by "the foolish people" on the back of an ass. This represents humanity's lack of understanding of the gifts of knowledge, and their relegating these to the domain of the merely animalistic. It is the Greek-mythic equivalent of "casting one's pearls before swine."

> The ass on his way home, being troubled with extreme thirst, came to a fountain; but a serpent, that was set to guard it, would not let him drink unless he gave in payment whatever that was that he carried on his back. The poor ass accepted the condition; and so for a mouthful of water the power of renewing youth was transferred from men to serpents.[20]

The story obviously parallels humanity's expulsion from Eden, namely, the denial of everlasting life for the crime of tasting forbidden fruit offered by a clever serpent. It is also reminiscent of the Biblical account of the hairy, animalistic Esau selling his precious birthright for a bit of red stew to his clever brother.[21] The moral is clear enough: bodily pleasure is the enemy of worldly

---

17. Spedding, Ellis, and Heath, *The Works of Francis Bacon*, VI:749. This is one of the many biblically inspired inversions in Bacon's essay. Here, Prometheus is a Jesus-like figure being condemned by the very crowd that he sacrificed himself for. Bacon takes the side of the crowd, not because Prometheus was wrong, but because the crowd's very complaining was necessary for their (i.e., humanity's) salvation.
18. Spedding, Ellis, and Heath, *The Works of Francis Bacon*, VI:747.
19. Spedding, Ellis, and Heath, *The Works of Francis Bacon*, VI:747.
20. Spedding, Ellis, and Heath, *The Works of Francis Bacon*, VI:745.
21. Gen. 25:30–34 NIV.

success, knowledge, and power. For Bacon, "knowledge that tendeth but to satisfaction, is but as a courtesan, which is for pleasure, and not for fruit or generation."[22]

A more democratic treatment of Prometheus is found in French Revolutionary art. Notably, the recurring figure of "Marianne" is emblematic of popular struggles against the *Ancien Régime* and feudal privilege. Its most famous manifestations are in Eugène Delacroix's painting *Liberty Leading the People* (1830), and later, as New York's "Lady Liberty," conceived by Frédéric Auguste Bartholdi. In each case, the feminine figure holds up a symbol of Olympian fire—either a bayoneted musket or a torch. The association between Marianne and Prometheus is so pronounced that an industrial city in eastern Ukraine even depicts the two figures side by side; each has a hand raised, holding a single torch together.

Here we have a thoroughgoing secularization of the mythic symbol. Prometheus now stands for political emancipation and struggle. But more than this, Prometheus himself becomes emancipated and fully autonomous. With Bacon's early modern depictions, Prometheus is still part of a basically legitimate Olympian order. Zeus is still wise; his rule still just. Prometheus's clever antics are integrated into this essentially harmonious worldview. Marianne, by contrast, is a revolutionary and not the loyal opposition.

Prometheus as revolutionary "Liberty" thus grates against the Olympian order. This is similar to the secularization of Prometheus in the hands of the poet Johann Wolfgang von Goethe. In his poem *Prometheus* (1789), Goethe's liberal and anticlerical sympathies are laid bare. Zeus and the other gods are now cast as heartless tyrants, wholly indifferent to human welfare and suffering. This is an indirect way of expressing his own pantheist sympathies as against traditional Christianity, especially the notions of a personal God and prayer. Goethe rejects the legitimacy of Olympian and divine rule as suffocating human potentialities and flourishing.

> I know of nothing poorer
> Under the sun than you gods.
> Wretchedly
> You feed your majesty
> On imposed sacrifices
> And the breath of prayers.

---

22. Francis Bacon, as quoted in Max Horkheimer and Theodor W. Adorno, *Dialectic of Enlightenment: Philosophical Fragments*, ed. Gunzelin Schmid Noerr, trans. Edmund Jephcott, Cultural Memory in the Present (Stanford, CA: Stanford University Press, 2002), 2.

You would waste away
If children and beggars
Were not hopeful fools [...]
Did you ever appease the pain
Of the sufferer?
Did you ever quench the tears
Of the fearful?[23]

This is not to say that Goethe was a populist. His Prometheanism is inherently aristocratic and heroic. The masses aren't capable, by their own faculties, of improving their condition, whether material or spiritual. Goethe's stance is thus liberal while also being antidemocratic. As he vehemently states, "Legislators and revolutionaries who promise equality and liberty at the same time, are either psychopaths or mountebanks."[24] The elevation of culture requires benevolent, tolerant monarchs in mutual support of genius artists and scientists. Goethe's dying words, *Mehr Licht!* (More Light!) should be associated with the spark of heroic genius rather than collective illumination.[25]

This was the positive side of Goethe's Prometheanism; he would explore its more pessimistic dimensions through his tragic play *Faust*.[26] In this work, Goethe describes Faust (another Promethean figure) as a late Renaissance scientist and polymath. In Faust's room, marked by high Gothic architecture, the main character commences with a dramatic lament:

I've studied now, to my regret,
Philosophy, Law, Medicine,
and—what is worst—Theology
from end to end with diligence.
Yet here I am, a wretched fool
and still no wiser than before [...]

---

23. Goethe's "Prometheus" as quoted in Joseph C. McLelland, *Prometheus Rebound: The Irony of Atheism* (Waterloo, ON: Wilfrid Laurier University Press, 1989), 25.
24. Goethe's *Maximen und Reflexionen* (1833) as cited by Erik von Kuehnelt-Leddihn, *Leftism Revisited: From de Sade and Marx to Hitler and Pol Pot* (Lanham, MD: Regnery Gateway, 1990), 9. This book, with a preface by William F. Buckley Jr., is just one example of contemporary reactionary thinkers citing Goethe as inspiration.
25. For a discussion of what might be Goethe's apocryphal last words, see Rüdiger Safranski, *Goethe: Life as a Work of Art*, trans. David Dollenmayer (New York: Liveright, 2017), 560.
26. Johann Wolfgang von Goethe, *Goethe's Collected Works, Volume 2: Faust I & II*, ed. and trans. Stuart Atkins (Princeton, NJ: Princeton University Press, 1994); references to Faust will be to the line number.

and find we can't have certitude.
This is too much for heart to bear!
I well may know more than all those dullards,
those doctors, teachers, officials, and priests,
be unbothered by scruples or doubts,
and fear neither hell nor its devils—
but I get no joy from anything, either.[27]

This is altogether different from the lament of the crowd we see in Bacon's Prometheus essay. In Bacon's still half-sacralized world (just at the beginning of the Scientific Revolution, and still before the Industrial), humans deride Prometheus's gift of knowledge because they only have the faintest idea as to what it means or how to use it. This is why they demand Zeus punish Prometheus, their would-be savior.

Faust, on the other hand, is distressed at having reached the end of natural knowledge. He has gone through all the books of the traditional disciplines and has learned everything there is to know. For all that, the result is a feeling of existential emptiness. All the knowledge of the world amounts to nothing substantial, apart, perhaps, from a sense of liberation from the strictures of religious morality.

The rest of the play centers around Faust's deal with the demon Mephistopheles. He sells his soul for magical and occult knowledge, granting him supernatural powers. The literary pattern once again repeats itself: man's hubris drives him to take the place of the gods. For "a brave man is not intimidated by celestial grandeur."[28]

Goethe takes a more nuanced view of hubris than does Bacon. The negation of divine bonds through a pact with the devil represents a moment in human progress. Even so, pride cometh before the Fall. In his Promethean enthusiasm, Faust decides to radically alter the entire landscape to his tastes, as if embarking on a post-Renaissance terraforming campaign. But completing this grand design means demolishing the small church and hut at the edge of his lands—ultimately leading to the demise of two elderly inhabitants.[29]

Rejuvenated by his pact with the devil, Faust directs his energies toward a young peasant girl, Gretchen. Through his lecherous advances, she becomes impregnated with Faust's bastard child, her reputation is ruined, and she is driven to insanity.

---

27. Goethe, *Goethe's Collected Works*, 354–70.
28. Goethe, *Goethe's Collected Works*, 712–13.
29. Goethe, *Goethe's Collected Works*, 11,160.

Salvation for Faust only comes in the last act of part II in the form of a spiritual essence in the guise of the Eternal Feminine.[30] The point for Goethe, it seems, is that the search for ever greater knowledge is a partial good, but must be tempered through a pantheistic recognition of Nature's authority. This is Goethe's way of squaring the circle, as it were. The masculine, irreverent, scientific drive for understanding cannot end merely with a conquest of nature, but instead, through a reconciliation with nature and a generalized compassion for all beings.

One may ask, however, whether Goethe's premises can truly lead to his conclusion? Is scientific knowledge perfected through universal compassion?

A close reading of the play seems to suggest not. Faust's original complaint was that scientific knowledge was exhausted; he defaults to an apophatic view that the sum of worldly learning amounts to nothing at all. But by the end of the play, the amoral and magical knowledge of Mephistopheles is also seen to be lacking. It must, finally, be supplemented with that which is beyond and outside of technical knowledge—the intuitive compassion of Mater Gloriosa, or the Eternal Feminine.

> Nature, mysterious in day's clear light,
> Lets none remove her veil,
> And what she won't discover to your understanding
> You can't extort from her with levers and with screws.[31]

The "feminine," therefore, may not represent the final perfection of knowledge after all, but is instead an external corrective.

> here is seen done;
> what's indescribable
> here becomes fact;
> Woman, eternally,
> shows us the way.[32]

For Goethe, the feminine is no advance. She only confirms Faust's original pessimism about knowledge. Faust cannot save himself through worldly means, and so an incomprehensible sense of grace is necessary. He does not

---

30. Goethe, *Goethe's Collected Works*, 12,110.
31. Goethe, *Goethe's Collected Works*, 672–75. This is a bit of foreshadowing from part I, referring to nature in general, but consistent with the character of Mater Gloriosa in part II.
32. Goethe, *Goethe's Collected Works*, 12,107–11.

attain ultimate unity with Nature as an extension of his own worldly strivings. Instead, this is the classic *"dea" ex machina* ending.

But these endings are unsatisfying precisely because they are bad storytelling. The resolution does not come about organically from the beginning and middle plot points; the natural light of reason is not intensified or perfected by the sacred feminine. On the contrary, the latter moderates and constrains the energies of the former. This is a wholly external sort of correction.

In the last act of the play, this is symbolized in very stark form: After causing the dispossession and death of the elderly couple, Faust is consumed with regret. At this point, the feminine spirit of Care manifests and blinds him. From that point on, Faust is remade in the image of altruism and selflessness, and ultimately his soul is saved from the devil.[33]

Whereas before Faust was motivated by a heroic egoism, he now marshals all the productive and technological forces of his commercial empire for humanitarian ends. He constructs a canal to make the area more habitable and productive, ordering Mephistopheles to mobilize his workforce.

> Use every means you can
> And get a plentiful supply of laborers;
> Use benefits and discipline to spur them on,
> Make payments, offer bonuses, conscript them!
> And day by day I want to be informed
> How the canal I've started is advancing.[34]

Faust becomes a martyr figure whose self-sacrifice is utterly disjointed from his original daring quest for knowledge and power. But this is a contorted Prometheanism which is now directed only toward the well-being of others. For his part, Mephistopheles is not impressed by this newfound moralism, mocking it as futile and banal. After all, every mortal is doomed to be reclaimed by the elements sooner or later. The general welfare is a lost cause.[35]

The German philosopher Friedrich Nietzsche appears to take Mephistopheles's side in his mockery of the humanitarian impulse. In *Human,*

---

33. Goethe, *Goethe's Collected Works*, 11,495–500. In this way, Goethe's feminine spirits function analogously to the androgynous Ghost of Christmas Past in Charles Dickens's *A Christmas Carol*, and even Cindy Lou Who and her singing compatriots in Dr. Seuss's *How the Grinch Stole Christmas!* Diverse as these works may be, what they have in common is an external, often feminine force which blunts the greed and misanthropy of the male protagonist.
34. Goethe, *Goethe's Collected Works*, 11,552–56.
35. Goethe, *Goethe's Collected Works*, 11,557–58.

All Too Human, Nietzsche derides the *Faust* story as typical of the romantic and tragic excesses of German literature.

> The Faust-Idea.—A little sempstress is seduced and plunged into despair: a great scholar of all the four Faculties is the evil-doer [...] Without the aid of the devil incarnate, the great scholar would never have achieved the deed [...] But for Goethe even this idea was too terrible [...] Even the great scholar [Faust], "the good man" with "the dark impulse," is brought into heaven in the nick of time [...] In heaven the lovers find themselves again.[36]

Nietzsche's sarcastic condemnation of the Faust story is in contrast to his general appreciation of Goethe himself. In particular, Nietzsche considered Goethe to be a "good European," given the latter's high level of culture and opposition to the levelling tendencies of democracy. But he disliked *Faust* because the story failed to reach its natural terminus. The world-conquering scholar, if worthy of the title, should not have let himself be weakened by effeminate impulses toward pity or Care.[37]

Nietzsche's early writings on Prometheus, in *The Birth of Tragedy*, confirm this view. As with Goethe, Nietzsche sees Prometheus as a tragic cultural hero representing the most sublime impulses of mankind. His theft of fire is a sacrilege; but this, says Nietzsche, is an "active sin." He immediately contrasts this with the biblical narrative of the Jews, where sinning is conceived as purely passive. In Genesis, Eve steals knowledge from God because of her curiosity, her ability to be deceived by the serpent, and above all, her "susceptibility to be led astray." For the Jews, surmises Nietzsche, evil has its origin in "eminently feminine feelings." According to Nietzsche, Prometheus is a far more fitting hero for the Aryan race.[38] He is not moved by impulses of the flesh or a passive, childlike curiosity. He sins with full presence of mind and intention. His challenge to the gods is open and masculine; it is a self-conscious act of mastery.

Here we see the central divergence between Nietzsche and Goethe. With Goethe, daring to know is eventually tempered by a care for mankind. Knowledge becomes tied to morality. For Nietzsche, what gives our daring

---

36. Friedrich Nietzsche, *Human, All Too Human: Parts One and Two*, trans. Helen Zimmern and Paul V. Cohn (Mineola, NY: Dover, 2012), 404.
37. Friedrich Nietzsche, *Twilight of the Idols*, trans. Duncan Large (New York: Oxford University Press, 1998), 55–58.
38. Friedrich Nietzsche, *The Birth of Tragedy*, trans. Douglas Smith (New York: Oxford University Press, 2000), 57.

to know value is the "daring" itself, unfettered by any external constraints—whether moral or material.

This distinction is tied up with Nietzsche's particular brand of atheism. For him, the old superstitions and religions may have been false, but they did serve a purpose. Pseudosciences like alchemy, astrology, and even witchcraft created "a thirst, a hunger, a taste for hidden and forbidden powers."[39] Even orthodox religions accomplished this in their own way. The very notion of a forbidden fruit of knowledge, and of everlasting life, stands to tempt us mortals. As with small children, we never would have desired such a thing before it was prohibited by a father figure.

But now, exclaims Nietzsche, God is dead.[40] We have killed God through our popular disbelief in revealed religion. In our modern, desacralized world, the theater lights have been raised and the curtains drawn back. We now realize that, all along, *we* were the ones who created knowledge. The harsh, forbidding God, too, was a chimera of our own, mortal imaginings—born of a psychological need to conceal our own shocking power from ourselves. For Nietzsche, Prometheus represents a humanity which has come of age and has seen that the world did not spring out of the head of Zeus after all, but is rather his own creation.

> Did Prometheus have to *fancy* first that he had *stolen* the light and then pay for that—before he finally discovered that he had created the light *by coveting the light* and that not only man but also the *god* was the work of his own hands and had been mere clay in his hands? All mere images of the maker—no less than the fancy, the theft, the Caucasus, the vulture, and the whole tragic *Prometheia* of all seekers after knowledge?[41]

A thoroughly desacralized world—lacking any stable architecture—is a nominalized, malleable, flux. *We* give it meaning and definition. But if God is dead, this does not mean, for Nietzsche, an end to authority or hierarchy. More than ever, the world needs its masters and creators. It cries out to certain strong souls to constitute it once again.

The key insight of the passage above is that Prometheus not only creates light (i.e., knowledge) and God, but also man himself. There is a striking, internal consistency here. In a world constituted by the will, there can be no sense of a stable and universal human nature that we all share, guaranteeing

---

39. Friedrich Nietzsche, *The Gay Science: With a Prelude in Rhymes and an Appendix of Songs*, trans. Walter Kaufmann (New York: Vintage Books, 1974), 240.
40. Nietzsche, *The Gay Science*, 181.
41. Nietzsche, *The Gay Science*, 240–41.

our equality. The self-creation of man will be the free act of certain cultural heroes and geniuses. It is not the prerogative of each individual to create him or herself, but the privilege of a worthy few to impose their designs on the rest. As Nietzsche unequivocally puts it, "My philosophy aims at an ordering of rank: not at individualistic morality."[42] Prometheus may represent a humanity which is fully in control; but not all people can be a Prometheus.[43]

The punishment and suffering of Prometheus, likewise, takes on a new meaning. The external penalty for stealing fire is nothing to him. For great souls are ever ready to suffer for their art and their free creations. The only thing which can wound a Prometheus is the thought that his greatness goes unrecognized—that his heroic acts are misattributed to the gods.

Misunderstood sufferers.—Magnificent characters suffer very differently from what their admirers imagine. They suffer most […] from their doubts about their own magnificence—not from the sacrifices and martyrdoms that their task demands from them. As long as Prometheus feels pity for men and sacrifices himself for them, he is happy and great; but when he becomes envious of Zeus and the homage paid to him by mortals, then he suffers.[44]

Thus, even the martyrdom and "self-sacrifice" of Prometheus are transformed in Nietzsche's hands. Stealing fire is not an act of selflessness for the sole benefit of humanity. It becomes, instead, aristocratic and prideful—the performance of *noblesse oblige*. The death of Prometheus is not Christ-like; it is not an act of solidarity with the suffering mass of humans. Nietzsche's cultural hero is above all that. The noble can, at most, have a high-handed sympathy for their social inferiors but never a sense of compassion or true identity. But even sympathy is incidental to the "activity" of the strong soul. Instead, Nietzsche's take on Prometheus is one of spectacular and prideful dying for art, a deed fitting for a tragic Aryan hero.

Here, it is worth pointing out that many liberal and Left-leaning scholars of Nietzsche will take extreme exception to our characterization of the scholar as a partisan of aristocracy, and especially, of an Aryan chauvinism. There is

---

42. Friedrich Nietzsche, *The Will to Power*, ed. Walter Kaufmann, trans. Walter Kaufmann and R. J. Hollingdale (New York: Vintage Books, 1968), 162.
43. For a criticism of Left and egalitarian appropriations of Nietzsche's philosophy, see Harrison Fluss's introduction to Domenico Losurdo, *Nietzsche, the Aristocratic Rebel: Intellectual Biography and Critical Balance-Sheet*, trans. Gregor Benton (Leiden: Brill, 2019), 1–13.
44. Nietzsche, *The Gay Science*, 216.

a long tradition of reading Nietzsche as a "liberal-at-heart," or even as simply an apolitical advocate of artistic free expression. It is, in any case, a well-known trait of academics to project their own ideological leanings onto a favored subject. "Tell me what you need and I will find you a Nietzsche quotation for it."[45]

Our point here, however, is that the Nietzschean ethos of artistic free expression is in no ways incompatible with an aristocratic ethos of mastery. To the contrary, the two are conceptually linked: Artistic freedom of the sort Nietzsche valorizes is not subject to any objective standards or democratic norms. Its very "freedom" involves a strong assertion of the will. And this, for Nietzsche, presumes a base of mediocrity, or an "underclass" that will act as the human scaffolding of rare genius.[46] Thus, even if Nietzsche did not have in mind the specific geneticist theories peculiar to twentieth-century Nazism, his notion of the Aryan race is every bit as wedded to the ideal of hierarchical "rank ordering."[47] The Aryan is noble, high-born, and spiritually free, and according to Nietzsche, these psychological traits are indeed correlated to physical features and racial blood ties.[48]

That Aryan masculinity is the organic basis for higher civilization is echoed in the pages of *Mein Kampf*. For Hitler, as with Nietzsche, Prometheus is Aryan.

> All human culture, all the results of art, science, and technology that we see before us today, are almost exclusively the creative product of the Aryan [...] [H]e alone was the founder of all higher humanity, therefore representing the prototype of all that we understand by the word "man." He is the Prometheus of mankind from whose bright forehead the divine spark of genius has sprung at all times, forever kindling anew that fire of knowledge which illumined the night of silent mysteries and thus caused man to climb the path to mastery over the other beings of this earth. Exclude him—and perhaps after a few thousand years darkness will again descend on the earth, human culture will pass, and the world turn to a desert.[49]

In the twentieth century, the ethos of heroic creation was taken up by the defenders of free-market capitalism. The writer Ayn Rand was instrumental

---

45. Losurdo, *Nietzsche, the Aristocratic Rebel*, viii.
46. Nietzsche, *The Will to Power*, 476; Friedrich Nietzsche, *On the Genealogy of Morals*, trans. Douglas Smith (Oxford: Oxford University Press, 1996), 3–38.
47. Frank Cameron and Don Dombowsky, eds., "Will to Power," in *Political Writings of Friedrich Nietzsche: An Edited Anthology* (London: Palgrave Macmillan, 2008), 276.
48. Nietzsche, *On the Genealogy of Morals*, 3–38.
49. Adolf Hitler, *Mein Kampf*, trans. James Murphy (London: Hurst and Blackett, 1939), 226.

in promoting this viewpoint as against the welfarism of liberals and social democrats, and against the "collectivism" of the Soviet Bloc. Rand's widely read works of fiction exposed the mass public to the Nietzschean celebration of the strong individual. Her writing is a constant warning against the dangers of social levelling and enforced compassion.[50]

One of Rand's earliest novels *Anthem* was also her most dystopian. Published in the doldrums of the late 1930s, it envisions a postapocalyptic future bereft of the most basic technologies. This hell-scape is dominated by an all-consuming state which methodically suppresses any sense of individualism.

Instead of family names, all citizens are given propagandistic monikers followed by a number (highly reminiscent of concentration camp victims). The novel's narrator and male protagonist is named Equality 7-2521; the female love-interest is called Liberty 5-3000. But by the end of the book, the main characters have triumphantly reconnected with their sense of individual self, predictably, under the banner of Prometheus.

> There was a time when each man had a name of his own to distinguish him from all other men. So let us choose our names. I have read of a man who lived many thousands of years ago, and of all the names in these books, his is the one I wish to bear. He took the light of the gods and he brought it to men, and he taught men to be gods. And he suffered for his deed as all bearers of light must suffer. His name was Prometheus.[51]

The newly minted "Prometheus" then chooses to name his partner "Gaea." In a particularly androcentric passage, he then lays out her role in the new, glorious future befitting her name: "Let this be your name, my Golden One, for you are to be the mother of a new kind of gods [*sic*]."[52] Prometheus, like his divine namesake, takes on a more active agenda. He will rediscover the lost technologies of a past golden age.

> I have learned that my power of the sky was known to men long ago; they called it Electricity. It was the power that moved their greatest inventions. It lit this house with light which came from those globes of glass on the walls. I have found the engine which produced this light.

---

50. Ayn Rand, while superficially critical of Nietzsche's irrationalism, was in fact influenced by his thought. See Jeff Walker, *The Ayn Rand Cult* (Chicago: Open Court, 2012), 277.
51. Ayn Rand, *Anthem* (New York: Signet, 1995), 98–99.
52. Rand, *Anthem*, 99.

> I shall learn how to repair it and how to make it work again. I shall learn how to use the wires which carry this power.[53]

The mythic parallels to the original Prometheus story of Hesiod are obvious: Knowledge is the thing that saves mankind from its doom, and technical knowledge is symbolized by captured fire, that is, electricity.

In Rand's novel, Prometheus's drive to rediscover technology is accompanied by an equally passionate impulse to separate himself from the rest of society. Not only will electricity furnish him and his family with creature comforts, but will also serve to protect him from the levelling forces of the surrounding culture.

> Then I shall build a barrier of wires around my home, and across the paths which lead to my home; a barrier light as a cobweb, more impassable than a wall of granite; a barrier my brothers will never be able to cross. For they have nothing to fight me with, save the brute force of their numbers. I have my mind.[54]

Rand's distinctly negative and libertarian conception of freedom is clear: "To be free, a man must be free of his brothers. That is freedom. That and nothing else."[55] And it is the mind of Rand's Prometheus—his ingenuity—which sets him apart from the physical coercion of the masses.

But just as with Nietzsche, mastery of oneself in isolation is not enough. It pertains to the nature of strong souls, not merely that they flourish individually, but that they come to lead other people. Rand's Prometheus builds for himself a citadel on a mountaintop. He intends to gather all those people who have not been completely evacuated of their individual spirit, and with these "fellow-builders" he resolves to "write the first chapter in the new history of man." Along with his biological son, these new men of history will once again learn to say the word "I." "And man will go on. Man, not men."[56]

The Promethean themes in Rand's writing continue through her later, more popular novels. In *The Fountainhead* (1943), the protagonist Howard Roark strikes the same individualist tone. A modernist architect (based partly on Frank Lloyd Wright), he has an uncompromising vision for his creations—one which is constantly challenged by the socialist architecture critic Ellsworth M. Toohey. Ultimately unable to realize his artistic intent, and not willing to

---

53. Rand, *Anthem*, 100.
54. Rand, *Anthem*, 100.
55. Rand, *Anthem*, 101.
56. Rand, *Anthem*, 104.

compromise, Roark chooses to destroy his creation (a New York housing project) with dynamite.

The *Fountainhead* contains an extended elegy for the Promethean spirit. This occurs during Roark's defense speech at his trial following the bombing.

> Thousands of years ago, the first man discovered how to make fire. He was probably burned at the stake he had taught his brothers to light. He was considered an evildoer who had dealt with a demon mankind dreaded. But thereafter men had fire to keep them warm, to cook their food, to light their caves. He had left them a gift they had not conceived and he had lifted darkness off the earth [...] Prometheus was chained to a rock and torn by vultures—because he had stolen the fire of the gods. Adam was condemned to suffer—because he had eaten the fruit of the tree of knowledge. Whatever the legend [...] mankind knew that its glory began with one and that that one paid for his courage.[57]

What is notable in Rand's treatment is the misanthropic and resentful quality of her Promethean characters. Not only will Howard Roark suffer for his art; he is willing to destroy for it as well. While Hesiod's Prometheus was disemboweled for helping mankind, Rand's twentieth-century Prometheus will not debase himself before second rate mediocrities and "moochers" in this way.

A still clearer expression of this tendency can be found in Rand's most widely read work, *Atlas Shrugged* (1957). The novel's hero is John Galt, an engineer and inventor who is resentful toward union bureaucracy and government interference with business. In response to a stultifying collectivism, Galt organizes businessmen, creative entrepreneurs, artists, and industrialists in a massive strike aimed at destroying bureaucratic society. In Rand's story, this capitalist strike has devastating effects on the civilization it leaves behind, now lacking the genius innovators and agents of progress who have always sustained the world. Not only is innovation gone, New York City can't even keep the lights on.[58]

> John Galt is Prometheus who changed his mind. After centuries of being torn by vultures in payment for having brought to men the fire of the gods, he broke his chains—and he withdrew his fire—until the day when men withdraw their vultures.[59]

---

57. Ayn Rand, *The Fountainhead* (New York: Bobbs-Merrill, 1971), 602.
58. Ayn Rand, *Atlas Shrugged* (New York: Signet, 1957), 703.
59. Rand, *Atlas Shrugged*, 478.

What sets this novel apart, however, is not the fate of the world without its industrial titans (a theme already explored in *Anthem*). Instead, *Atlas Shrugged* is most revealing in its utopianism. The new society, Galt's Gulch, named after its founder, is an idyllic version of the American West set in Colorado. It consists solely of those individuals whom Rand considers to be "producers." These include farmers, bankers, mechanics, inventors, and other small business owners. It excludes corporate functionaries and lawyers and, of course, organized labor.[60] In fact, the Gulch operates without a dedicated workforce whatsoever. The high standard of living (evidenced by superior technology and the lack of poverty or crime) seems to materialize mostly through the sublime intellectual efforts of its inhabitants. This is a continuation of the Promethean theme, where intellectual prowess displaces the need for physical work and material production.

Even reproduction seems to occur immaculately. The Gulch is devoid of nearly any named women as residents, and there is a pronounced hostility to family dependency. Ironically, this is most stridently expressed by the sole, maternal figure in this society (a professional baker with two children of her own):

> You know, of course, that there can be no collective commitments in this valley
> and that families or relatives are not allowed to come here, unless each person takes the striker's oath by his own independent conviction.[61]

It is unclear if the requirement to take this oath extends to her own children, aged four and seven.

Both the relative absence of women and a working class indicate the persistent hostility toward material, social bonds. This repeats the androcentrism of *Anthem* wherein Gaea is subservient to her husband Prometheus. What is fascinating about *Atlas Shrugged* is that, except for the sincere ideological convictions of its author, it would read as a sharply sarcastic piece of farce. The very notion of a "capitalist strike" devoid of workers suggests not plentiful utopia but starvation and plague. Who will pick the crops, take out the trash, maintain machinery, and the like?

The only reason why Galt's Gulch seems to function at all is that Rand stacks the deck. Absent a dedicated working class, each of the small business owners is assumed to perform their own manual labor, or to casually subcontract such

---

60. Lisa Duggan, *Mean Girl: Ayn Rand and the Culture of Greed* (Oakland: University of California Press, 2019), 61.
61. Rand, *Atlas Shrugged*, 720.

labor with one another on a face-to-face basis. But the advanced technology of the Gulch, and its modern standard of living, are clearly incongruous with such homespun, premodern social relations.

The Gulch is hardly capitalist, except as an idealized version of capitalism's very early prehistory.[62] While it is populated by bankers, mechanics, electricians, and other modern professionals, the overall social structure is the Jeffersonian ideal of the independent yeoman farmer. But now, absurdly, there is also the yeoman banker and yeoman electrical engineer as well! How there could be bankers without banks, and modern banking without surplus value and profits, and profits without a mass of exploited workers, is never explained; nor can it be. Neither is how there could be specialized professions such as an electrician or plumber without a persistent division of labor, as though these were merely artisanal choices.

The blinkered utopianism of Rand is well criticized by the words of Karl Marx about a century before the publication of *Atlas Shrugged*. Against the utopians of his day, Marx polemicized that they "want all the advantages of modern social conditions without the struggles and dangers necessarily resulting therefrom. They desire the existing state of society, minus its revolutionary and disintegrating elements. They wish for a bourgeoisie without a proletariat."[63] But this frictionless, matter-less fantasy of Rand represents the apogee of Promethean thought in late capitalism. This is necessarily "neoreactionary" in that it imagines an advanced, technology-infused future atop the social relations of a premodern world.

Given these contradictions, not everyone will be so sanguine about a Promethean brave new world. The promise of advanced technologies has often been tempered by a fear of overreach; natural and social limits are there for a reason. Directly competing with Rand's capitalist heroism is the literary tradition centered upon the mad scientist: The pessimistic answer to John Galt is Victor Frankenstein.

The subtitle of Mary Shelley's seminal work, *Frankenstein*, is "The Modern Prometheus," though this phrase predates Shelley herself. "Modern Prometheus" was, for example, used by Immanuel Kant to describe Benjamin

---

62. This, of course, minus the idea of original accumulation. Notably, there is no indigenous population or former inhabitants of the land. For a discussion of this omission, see Duggan, *Mean Girl*, 61.
63. Karl Marx and Friedrich Engels, "Manifesto of the Communist Party," in *Marx & Engels: Collected Works*, Volume 6 (London: Lawrence and Wishart, 2010), 513. Marx criticizes Proudhon's image of Prometheus as divorced from class struggle and material relations, that is, as an etherealized reason apart from actual humanity. See Karl Marx, "The Poverty of Philosophy," in Marx and Engels, *Marx & Engels*, 57–59.

Franklin and his 1752 experiments harnessing lightning. Sarcastically, the Earl of Shaftesbury also used the appellation "modern Prometheus" to deride alchemists and other pseudo-scientists of his day.[64] But the proximate influence on Shelley's Prometheus likely came from her father, the philosopher William Godwin, who wrote extensively on the mythic figure. It is to Godwin, himself, that Shelley's *Frankenstein* is dedicated.[65]

While the Frankenstein image predates both Rand and late capitalism, it certainly has impressive staying power. In the twentieth and twenty-first centuries, those skeptical of modern science have used the "Frankenstein" moniker as a means of derision, especially in the popular press. Genetically modified crops are called "Frankenfoods"; genetic engineering is labeled a "Frankenstein" discipline; animal cells integrated into the human body are "Franken-organs"; the possibility of cloned pets are disparaged as "Frankenpets"; even nuclear weapons have been described as "Frankenstein monsters."[66] The father of the atomic bomb, J. Robert Oppenheimer, has tellingly been praised *both* as the "American Prometheus" and chided as the "20th Century Frankenstein."[67]

In Shelley's novel, the scientist Victor Frankenstein separates himself from society in order to fashion his artificial man. He engages in a number of taboo activities including the collection of body parts from morgues and cemeteries. No thought is given to the traditional strictures of polite Christian society.

---

64. Shaftesbury's *The Moralists* as cited in editorial notes to Mary Shelley, *Frankenstein or The Modern Prometheus*, ed. Maurice Hindle (London: Penguin Classics, 2003), xlix.
65. Mary Shelley's reception of the Prometheus myth was shaped by her father's translation of Ovid, rather than Hesiod. For a discussion of this, see the Introduction to the 1818 edition of the text. Mary Shelley, *Frankenstein or the Modern Prometheus*, ed. Nick Groom (Oxford: Oxford University Press, 2018), xxix.
66. Henry Miller and Gregory P. Conko, *The Frankenfood Myth: How Protest and Politics Threaten the Biotech Revolution* (Westport, CT: Praeger, 2004); George A. Hudock, "Gene Therapy and Genetic Engineering: Frankenstein Is Still a Myth, but It Should Be Reread Periodically," *Indiana Law Journal* 48, no. 4 (1973): 533–58; Liz Stinson, "Imagine If a Designer Could Make You a Custom, Frankenstein Heart," *Wired*, August 1, 2013, https://www.wired.com/2013/08/are-these-frankenstein-organs-the-implants-of-the-future/; "Frankenpet," *Perspectives* (San Francisco, CA: KQED, November 24, 2019), https://www.kqed.org/perspectives/201601139297/frankenpet.
67. Kai Bird and Martin J. Sherwin, *American Prometheus: The Triumph and Tragedy of J. Robert Oppenheimer* (New York: Random House, 2005), 293. For an essay comparing the personalities and life trajectories of Oppenheimer and Victor Frankenstein, see Leonard Isaacs, "Creation and Responsibility in Science: Some Lessons from the Modern Prometheus," in *Creativity and the Imagination: Case Studies from the Classical Age to the Twentieth Century*, Volume 3 (Newark, NJ: University of Delaware Press, 1987), 59–104.

Darkness had no effect upon my fancy, and a churchyard was to me merely the receptacle of bodies deprived of life, which, from being the seat of beauty and strength, had become food for the worm. Now I was led to examine the cause and progress of this decay and forced to spend days and nights in vaults and charnel houses.[68]

Frankenstein represents an alternate ideological path, one not taken by the other modern writers on Prometheus. Goethe's *Faust* was the decision-point, the cultural watershed where one had to decide whether the Promethean figure was heroic or villainous. As we saw, this play includes two, unassimilable elements: On the one hand, the intrepid scientist who flouts the bounds of "normal" science and morality, and on the other hand, the saving power of Care and a selfless altruism. But there is no dialectical reconciliation between these two elements—only an artificial, *dea ex machina* ending. The play, in its contrived, papering over of contradictions calls out for a decision to be made: praise the daring scientist after all, or reject his hubristic project from the beginning.

While writers like Nietzsche and Rand resolve this question by doubling down on an individualistic, antisocial heroism, Shelley goes in the exact opposite direction, namely, condemnation of scientific excess. Her Frankenstein begins just as Faust did—a bored scholar, chafing under the yoke of academic conformity and craving something more. "I was required to exchange chimeras of boundless grandeur for realities of little worth."[69] Like Faust, Dr. Frankenstein exchanges "normal," respectable science for the occult arts—famously, the ability to create life out of dead matter with the Promethean fire of electricity.

Initially, he sets out to create an entire race of men, mirroring the heroic acts of the titan himself. "A new species would bless me as its creator and source [...] No father could claim the gratitude of his child so completely as I should deserve theirs."[70] But this Promethean project is soon aborted as Dr. Frankenstein is horrified by the very first of his creations which he deems monstrous. He refuses to fashion a female companion for the monster, lest a "race of devils" be loosed upon the earth.[71]

Like Uranus shoving his progeny back into the Earth or Cronus consuming them, Victor's blunting of creation is akin to what we saw in the Hesiodic

---

68. Shelley, *Frankenstein or The Modern Prometheus* (2003), 52–53.
69. Shelley, *Frankenstein or The Modern Prometheus* (2003), 48.
70. Shelley, *Frankenstein or The Modern Prometheus* (2003), 55.
71. Shelley, *Frankenstein or The Modern Prometheus* (2003), 170.

myths. It represents a psychological fear of being displaced by one's own children.

What is truly horrifying about Frankenstein's monster is his uncanny freedom and independence. He learns to read, develops a conception of himself, and even asserts his own needs and right to happiness. Hitting up against a world which denies this inner subjectivity, the monster rebels violently, even killing Victor's fiancée.[72]

Dr. Frankenstein pursues his hated creation to the ends of the Earth, hoping to kill it and rectify his mistake. But unlike the *Faust* narrative, the mad scientist is never redeemed. He fails in his mission to kill the monster and never puts his ingenuity to humanitarian use. Instead, defeated and weak, his dying words are an elegy for the quiet, tranquil life and a biting indictment of youthful ambition.[73]

In crafting this sort of morality tale, Shelley eschews the Aryan hero-worship of Nietzsche and the capitalist misanthropy of Rand. But for all that, there is a conservative streak running through this work of gothic fiction. As many commentators have pointed out, Frankenstein's monster can be read as modernity personified. Writing in the midst of the First Industrial Revolution, Shelley's novel is an extended meditation on the birth of the modern working class; it is also a warning to the ruling elite.[74] Just as Dr. Frankenstein animates his creature, so the bourgeoisie calls the proletariat into existence, and assumes mastery over it.

But this creation—armed with the awesome forces of mechanized production—threatens the supremacy of the capitalist class. As Marx and Engels put it in the *Communist Manifesto*, "the bourgeoisie forged the weapons that bring death to itself; it has also called into existence the men who are to wield those weapons—the modern working class—the proletarians."[75] This is the old tale of *The Sorcerer's Apprentice* (1797), another Goethe classic, but with a pronounced, sociopolitical edge. For the rebellion of creation now represents the overturning of established relations.

The question remains, however, as to what should be the moral remedy for Dr. Frankenstein's excesses? In the case of the "modern Prometheus," Mary

---

72. Shelley, *Frankenstein or The Modern Prometheus* (2003), 199.
73. Shelley, *Frankenstein or The Modern Prometheus* (2003), 220. While Victor's final speech is full of regret, it nonetheless includes a moment of ambiguity, leaving open the possibility that "another may succeed" where he failed.
74. Franco Moretti, *Signs Taken for Wonders: On the Sociology of Literary Forms* (London: Verso, 2005), 83–109; David McNally, *Monsters of the Market: Zombies, Vampires, and Global Capitalism* (Leiden: Brill, 2011), 17–88.
75. Marx and Engels, "Manifesto of the Communist Party," 490.

Shelley is on the side of Zeus, or at least, what he represents. Especially in Hesiod's telling, Prometheus is rebuked by the Olympian gods for his arrogance and pride. When Hesiod then offers practical advice to the reader, it similarly consists of a praise for tranquil village life, moderation, and the traditional family. Victor Frankenstein is heedless of such parochial advice and pays dearly for it. His path of self-destruction starts, not at the creation of his monster, but earlier when he leaves the comfort of his family home for an ambitious scientific career in a foreign university town.[76] Shelley's pessimism about progress demands respect for traditional family relations, religious mores, and thus, an "Olympian" sense of social order and moderation.[77]

The Gaian Alternative

Nonetheless, in this mythic complex, another alternative presents itself. To remedy the ambitious excesses of a Dr. Frankenstein, one may return to Olympus, or more radically, seek salvation in Nature herself. Those horrified by techno-scientific rebellion, and underwhelmed by the conservative status quo, will opt for the Gaian alternative. When speed, novelty, and innovation threaten to break the world apart and profane all that is holy, Gaia must reassert herself and rein in the rebellious human spirit.

Here there is an interesting reversal of the mythic and historical chronologies. In Hesiod (as with Aeschylus and Ovid), Gaia is the primordial being who precedes both Zeus and Prometheus. Nevertheless, the "Gaian movement" occurred very recently, postdating both the Scientific and Industrial Revolutions. Gaianism becomes a popular tendency only in the twentieth century.[78] It was never a preindustrial ideology, but instead a reaction to the perceived excesses of human productivity and technology. With the backdrop of the atomic bomb, rapid increases in population, the overuse

---

76. Shelley, *Frankenstein or The Modern Prometheus* (2003), 269, The university town of Ingolstadt is the birthplace of the Bavarian order of the Illuminati. The Illuminati has long been a signifier, especially among traditionalist writers, for subversive conspiracy and revolution.
77. For a discussion of possible affinities between Shelley and Burkean conservatism, see Adriana Craciun, "Frankenstein's Politics," in *The Cambridge Companion to Frankenstein* (Cambridge: Cambridge University Press, 2016), 95.
78. NB: We speak of "Gaianism," not as a top-down political movement, but rather as a constellation of ideas and political trends. Furthermore, we recognize that romantic ecologies predate the twenty-first century. Nonetheless, the emergence of the scientific "Gaia hypothesis" along with the modern "feminist spirituality" movement are unique to our era. For a discussion of earlier romantic venerations of Nature, see Jonathan Bate, *Romantic Ecology: Wordsworth and the Environmental Tradition* (New York: Routledge, 1991).

of pesticides in industrial agriculture, and dozens of other real or perceived threats, "Gaianism" became an intellectual force of its own.[79]

But Gaianism was, as well, a response to more fundamental, structural changes happening in the economy after the postwar boom of the 1950s. Popular Gaian ideologies were as much about the environment as they were about the changing role of women in society. A series of recessions took place during the periods of 1957–58 and 1973–75 (here, elicited by the oil crisis), and then again in 1981–82.[80] Given these crises, many middle-class families, which had previously depended only on male income, found themselves requiring two paychecks in order to maintain their expected lifestyles and levels of consumption.[81] This, of course, had both a liberating and exploitative effect. Women were no longer entirely dependent on husbands and fathers for their livelihoods, but now were compelled to sell their labor-power for a wage.

This placed many female workers in an impossible situation. They were compelled to enter the paid workforce because of economic necessity. At the same time, a conservative reaction saw female paid labor as a dangerous social trend. It was contended that women in the workplace would lead to a breakdown of the traditional family, neglected children, and an overthrow of male authority. We see such anxieties across the globe, well expressed by Pat Robertson's Christian conservatism, no less than Sayyid Qutb's earlier Right-wing Islamism:

> The feminist agenda is not about equal rights for women. It is about a socialist, anti-family political movement that encourages women to leave their husbands, kill their children, practice witchcraft, destroy capitalism and become lesbians.[82]

> In the Islamic system of life [...] family provides the environment under which human values and morals develop and grow in the new generation

---

79. Lawrence E. Joseph, *Gaia: The Growth of an Idea* (New York: St. Martin's, 1990).
80. For an extended discussion on the breakdown of the Keynesian consensus, as a result of these crises, see Geoffrey Pilling, *The Crisis of Keynesian Economics: A Marxist View* (London: Croom Helm, 1987).
81. Of course, this picture is complicated by the realities of racial disparity in the United States. Since at least 1870, half of African American women participated in the labor market as compared to only 16.5 percent of their white counterparts. For a discussion of this, see Asha Banerjee and Cameron Johnson, "African American Workers Built America," *Center for Law and Social Policy* (blog), February 26, 2020, https://www.clasp.org/blog/african-american-workers-built-america.
82. Pat Robertson's fundraising letter on behalf of the Christian Coalition as cited in the Associated Press, "Robertson Letter Attacks Feminists," *New York Times*, August 26, 1992, sec. A.

[…] [I]f woman's role is merely to be attractive, sexy and flirtatious, and if woman is freed from her basic responsibility of bringing up children; and if […] she prefers to become a hostess or a stewardess in a hotel or ship or air company […] rather than in the training of human beings […] then such a civilization is "backward" from the human point of view, or "jahili" [ignorant] in the Islamic terminology.[83]

Of course, it is the logic of capitalism which compels women to enter the workplace, and not some demonic drive to destroy the traditional family. If the need to work allowed some women to escape the stultifying isolation of domestic life, the private firm presented them with its own share of authoritarianism and control.

Constantly rebuffed by entrenched patriarchal interests, many have turned to a middle-class ideology of their own: The problem is not the workplace as such, but the *toxic male* workplace. The problem is not the exploitation of labor under capitalism, but cutthroat forms of *male-dominated* competition. If we are to refashion both the economy and our culture in a more humane, inclusive, and tolerant vein, then we must valorize those virtues which have traditionally been associated with the feminine. The vision is not one of overthrowing all social relations, but instead of making them more humane.[84] Capitalism must become more caring, more environmentally conscious, more generous, and above all, smaller.

The idea that capitalism should be overthrown by an international working-class movement is dismissed as either implausible or undesirable. For this would only repeat the totalizing and megalomaniacal tendencies of the current ruling elites. And so within second-wave feminism and the Gaian countercultures beginning in the 1970s, we see an ethos of "small is beautiful" and of "treading lightly upon the earth."[85]

There is here a constellation of belief which includes an affirmation of community, environment, and gender equality as against a bounded individualism,

---

83. Sayyid Qutb, *Milestones* (New Delhi: Islamic Book Service, 2006), 66–67.
84. For a recent example of this thinking, see Celia V. Harquail, *Feminism: A Key Idea for Business and Society* (New York: Routledge, 2020). Harquail advocates, "practicing feminist values to make businesses more successful and more just."
85. E. F. Schumacher, *Small Is Beautiful: Economics as If People Mattered* (New York: Harper Perennial, 1973), 4, 68. Schumacher complains that Marx (and by extension Marxism) commits the sin of anthropocentrism by equating all history with "class struggle," and all value with human labor. For an ecofeminist critique of Marx as instrumentalizing nature, see Ariel Salleh, "Sustaining Marx or Sustaining Nature? An Ecofeminist Response to Foster and Burkett," *Organization & Environment* 14, no. 4 (December 2001): 443–50.

human exceptionalism, and patriarchy. The ego is associated with the typically overambitious male, who steps over all competition and instrumentalizes Nature for his own narrow ends.[86] The ego must be humbled in favor of the community and a respect for the Earth which is always beyond full comprehension or instrumentalization. Not everything or everyone is a resource to be exploited or an adversary to be overcome. Gaianism, at least in this early phase, accommodates itself well to a green and welfare-oriented capitalism.

To be sure, a more recent wave of Gaian thought has staked out a position which they see as more radically anti-capitalist.[87] The examination of these figures is the central object of Chapter 3 of this book. However, before we can analyze these latter-day Gaians, it will be productive to first examine the origins of this movement.

Gaian Spirituality

The Goddess movement, sometimes called the feminist spirituality movement, has its origins in archeology. This is because its fundamental premise is that the patriarchy we find ourselves in today is neither universal, nor permanent, nor natural. As such, proponents of the Goddess movement often make the case for reforming contemporary society by invoking a glorious golden age of gender equality, usually located in prehistoric times.

But one cannot speak of archeology's impact on this brand of feminism without reference to the works of Marija Gimbutas. A Lithuanian émigré who fled the advance of Stalin's forces during World War II, Gimbutas eventually became a prominent archeologist in the United States. In her early work, she is rightly credited for developing the "Kurgan hypothesis" in the 1950s. This theory, still widely accepted today, gives an account for the spread of Indo-European languages throughout much of Europe and regions of South Asia. According to Gimbutas, the Proto-Indo-European (PIE) people resided, originally, on the Pontic steppe, adjacent to the Black Sea. They were a seminomadic, pastoral people who bred horses and were, above all else, highly militaristic. Since this theory refers to a period of prehistory which was largely preliterate, the "Kurgans" are given this name by

---

86. Exemplary of this standpoint is the diagnosis of "isms of domination" in K. J. Warren, "The Power and the Promise of Ecological Feminism," *Environmental Ethics* 12, no. 2 (1990): 125–46.
87. See, for example, Donna J. Haraway, *Staying with the Trouble: Making Kin in the Chthulucene* (Durham, NC: Duke University Press, 2016). The work of Donna Haraway will be analyzed at length in Chapter 3.

Gimbutas because of the large burial mounds (i.e., *kurgans*) endemic to their society.[88]

Linguistic studies give credence to the Kurgan hypothesis, noting strong linguistic resemblances between various Eurasian languages and the putative language of the Proto-Indo-Europeans. For example, the PIE meaning of the word "axle" (central to their horse-driven, martial culture) is retained in a diverse array of Indo-Iranian, Balto-Slavic, Germanic, Celtic, Italic, and Greek languages.[89]

Other likely cognates include the words for "father," "mother," "wood," and "water." Perhaps more impressively, recent genetic studies seem to confirm a distribution of genes from the area around the Pontic steppe. A 2015 study showed that around 3500 BCE, approximately 75 percent of the human gene pool in Europe was likely replaced by those found among the Yamnaya people (associated with the Kurgan homeland). This was roughly the same time period that the Kurgan hypothesis places PIE migrations to Europe. Other recent genetic studies confirm, while complicating, this hypothesis—showing that a migration did likely occur, but probably on a smaller scale and with some of the population returning to the steppes in a pendulum-type movement.[90]

The Goddess movement predated Gimbutas's work, but once her archeological contributions were popularized, some "feminist spiritualists" adopted it with an uncommon passion.[91] This is largely because Gimbutas not only advanced an empirical thesis about Neolithic cultural diffusion, but a normative thesis as well. This is particularly evident in her later works, where reciprocally, Gimbutas began to affirm and endorse aspects of the Goddess movement herself.[92] For her, the Proto-Indo-Europeans were not merely one culture among many; they were the veritable standard-bearers of patriarchy and violence. The culture which they displaced, what Gimbutas called "Old Europe," was by contrast the paragon of peace and civility—a rich society of plenty where women were held in high esteem and the Goddess was

---

88. Marija Gimbutas, "The Indo-Europeanization of Europe: The Intrusion of Steppe Pastoralists from South Russia and the Transformation of Old Europe," *Word* 44, no. 2 (1993): 205–22.
89. David W. Anthony and Don Ringe, "The Indo-European Homeland from Linguistic and Archaeological Perspectives," *Annual Review of Linguistics* 1 (2015): 199–219.
90. Roni Jacobson, "New Evidence Fuels Debate over the Origin of Modern Languages," *Scientific American*, March 1, 2018, https://www.scientificamerican.com/article/new-evidence-fuels-debate-over-the-origin-of-modern-languages/.
91. We use the label "Goddess movement" in a generic way to refer to the diverse array of spiritual feminists in the mid-to-late twentieth century and into the twenty-first.
92. Marija Gimbutas, *The Language of the Goddess* (New York: Harper Collins, 1991), xxi.

worshipped. For the feminist spirituality movement, these supposed findings engendered the hope that, if only our mythic orientation could be corrected, then social, political, and environmental crises could finally be tackled and perhaps overcome.

The specific nature of the Goddess religion, as envisioned by Gimbutas, is critically important for its role in inspiring some feminist spiritualists. First, the Goddess of Old Europe is said to be a prime "Creatrix." She is not a mere fertility deity, the wife or consort of a male god, nor a symbol of male erotic desire.[93] The Neolithic Goddess only becomes closely associated with fertility with the advent of agricultural food production.[94] She is the personification of vital Nature itself.

> Indeed, she is earth fertility incarnate: moist, mysterious, strong. She is pure and immaculate, creating life from herself, from her moist womb. She continually performs the miracle of magical transformation. Everything born from the earth is brimming with the life force. Flower, tree, stone, hill, human, and animal alike are born from the earth, and all possess her strength. Sacred groves, meadows, rivers, leafy trees, and gnarled, contorted trees growing together from several stumps are particularly charged with the mystery of life. The Earth Mother creates a cover for the earth that is lush, blossoming, and enchanted.[95]

The Goddess' mythic functions are said to represent the three primary functions of the Earth: life, death, and regeneration. These, likewise, mirror the three phases of the moon as understood in antiquity (new, waxing, and old), and are expressed in the female personae of the Maiden, the Nymph, and the Crone.[96]

The Goddess, in her seemingly contradictory phases, actually represents the vitalistic flux of Nature itself, which is both life and death dealing. The Maiden represents the virginal potential for life, the Nymph the actual generation of life, and the Crone the final phase of life (i.e., death) which then elicits a rebirth just as the cycles of the moon repeat themselves. Archeologically, these alleged insights are organized under the maxim that "the tomb is womb."[97]

---

93. Marija Gimbutas, "The Earth Fertility of Old Europe," *Dialogues d'histoire Ancienne* 13 (1987): 11.
94. Gimbutas, *The Language of the Goddess*, 316.
95. Gimbutas, "The Earth Fertility of Old Europe," 23.
96. Gimbutas, "The Earth Fertility of Old Europe," 11,12; Gimbutas, *The Language of the Goddess*, 316.
97. Gimbutas, *The Language of the Goddess*, 20.

In other words, the Goddess always calls back her children at the end of their terrestrial sojourns, only to be born again in the next cycle of regeneration.

The effect of Gimbutas's archeological work within certain domains of feminism is clear. A notable example are the writings of Charlene Spretnak who has advanced the cause of ecofeminism, and whose book *Green Politics* (1984) helped to intellectually found the green political movement in the United States.[98]

In defending Gimbutas against her critics, Spretnak repeats the former's central claim that primal religion must have been gynocentric because of the obvious parallels between the phases of life on earth, on the one hand, and the rhythms of the female body, on the other.

> Moreover, in many early cultures around the world the powers of nature were perceived metaphysically to have female qualities, presumably because of the easily observed parallels: women have a red tide that flows in rhythm with the cycles of the moon; they can swell up like the full moon; and they can bountifully produce (babies and milk), as does nature.[99]

On this view, women's bodies are uniquely the microcosms of the universe itself.

Gimbutas and her followers see the Neolithic worship of female figures as aspects of a singular and universal Goddess.[100] Such figures were not, according to Gimbutas, merely local deities, particular to a specific aspect of nature, a stream, a river, a rock, or even a geographical region. In her telling, all of Europe (and perhaps beyond) represented a consistent Goddess culture, worshipping an equally unique and unitary Creatrix.

Further, male divinity in Old Europe was said to be entirely subordinate to this supreme Goddess. Here one must be careful with language, for Gimbutas and her followers are often reluctant to speak of subordination of men to women in their utopian vision of Neolithic and Copper Age Europe. This would merely reproduce the "unbalanced" gender domination of patriarchy and androcracy after the Kurgan invasions.

---

98. Fritjof Capra and Charlene Spretnak, *Green Politics: The Global Promise* (New York: E.P. Dutton, 1984).
99. Charlene Spretnak, "Anatomy of a Backlash: Concerning the Work of Marija Gimbutas," *Journal of Archaeomythology* 7 (2011): 38.
100. Gimbutas, *The Language of the Goddess*, 316.

In Old Europe the world of myth was not polarized into female and male as it was among the Indo-European and many other nomadic and pastoral peoples of the steppes. Both principles were manifest side-by-side. The male divinity in the shape of a young man or a male animal appears to affirm and strengthen the forces of the creative and active female. Neither is subordinate to the other; by complementing one another, their power is doubled.[101]

Still, it is clear that Gimbutas's view was that "maleness" was peripheral within Old European spirituality. She claims, for example, that there is no trace of a father figure in Paleolithic excavations.[102] Gimbutas does admit that later figures, discovered in other excavations, do depict male persons, male animals, and even phalluses. But instead of revising her Goddess hypothesis, these findings are merely reinterpreted to fit the original theory. Male figures are claimed to represent the spontaneous generation of life which aids and enhances the prime "Goddess Creatrix."

The upshot of this archaeomythology is not so much its scientific value as its polemical use for today's politics. The point of "discovering" and describing such an idyllic past is to prove that there is a model for reforming the present systems of injustice, environmental degradation, and violence (surely worthy goals in themselves). To that end, Gimbutas connects the alleged mythologies of Old Europe with its pristinely humane politics and advanced culture.

This culture took keen delight in the natural wonders of *this* world. Its people did not produce lethal weapons or build forts in inaccessible places, as their [Indo-European] successors did, even when they were acquainted with metallurgy. Instead, they built magnificent tomb-shrines and temples, comfortable houses in moderately sized villages, and created superb pottery and sculptures. This was a long-lasting period of remarkable creativity and stability, an age free of strife.[103]

If one defines civilization as the ability of a given people to adjust to its environment and to develop adequate arts, technology, script, and social relationships it is evident that Old Europe achieved a marked degree of success.[104]

---

101. Marija Gimbutas, *The Goddesses and Gods of Old Europe: Myths and Cult Images*, New and Updated Edition (Berkeley: University of California Press, 1982), 237.
102. Gimbutas, *The Language of the Goddess*, 316.
103. Gimbutas, *The Language of the Goddess*, 321.
104. Gimbutas, *The Goddesses and Gods of Old Europe*, 17.

The later Kurgan invasions are depicted as the artificial, brutal interruption of a flourishing and ancient civilization which is morally superior to our own. The political task now is, therefore, not the creation of some novel, as yet untried, experiment in gender equality; it is rather to set right that natural, gender-balanced rule of women. For Gimbutas and her devotees, female rule is something to be restored, not invented.

For all of the reluctance of Gimbutas to speak of female supremacy or matriarchy, this is nonetheless revealed to be her normative position. She frequently valorizes Old Europe as having a pantheon which "reflects a society dominated by the mother."[105] Elsewhere, she writes approvingly how "the culture called *Old Europe* was characterized by a dominance of women in society and worship of a Goddess." By contrast, men (in spirituality as in politics) are mere adjuncts. They serve a "spontaneous and life-stimulating" function, in effect helping the Goddess Creatrix carry out her own vital work.[106]

With the Kurgan invasions, claims Gimbutas, there was not a wholesale displacement of the Goddess religion, but rather a subordination to the Indo-European, androcentric belief-system. Goddess religion was either pushed underground, confined to the domestic sphere, or else integrated into the Indo-European pantheon, where Old European goddesses were "married" to Indo-European, masculine gods.[107] Hardly an egalitarian fusion, the Old European goddess figures were reinterpreted as objects of beauty and male desire, and occasionally, took on stereotypically masculine characteristics themselves. Gimbutas, for example, speaks of the Greek Athena as no longer evoking aspects of the natural world, but instead representing warfare.[108]

In a parallel way, some in the feminist spirituality movement draw on theories that Canaanite religion (prior to monotheistic Judaism) was also matrifocal, and that this was accompanied by a culture which was matrilineal, if not entirely matriarchal. Of particular interest is the figure of "Asherah" ("Athirat" in Ugaritic) who was likely described as the "Queen of Heaven" and associated with sacred trees or groves. But with the consolidation of Hebrew monotheism, a violent suppression of both Asherah and women's own power was imposed.

In the case of ancient Israel, the feminine was suppressed even more completely as compared to the Indo-European context. For given the premise of monotheism, the goddesses could not be married to the dominant male gods, but instead, had to be discarded altogether. And so we have the

---

105. Gimbutas, *The Goddesses and Gods of Old Europe*, 237.
106. Gimbutas, *The Goddesses and Gods of Old Europe*, 9–10.
107. Gimbutas, *The Language of the Goddess*, 318.
108. Gimbutas, *The Language of the Goddess*, 318.

injunction in Judges 6:25 for the devotional pillars of Asherah to be torn down. Monotheism's intolerance of independent female divinity is taken as an indictment of modern established religions in general, especially those of the Western Abrahamic variety. As Merlin Stone, the author of *When God Was a Woman*, put it, "Most modern religions—particularly Judaism, Christianity and Islam—believe in a masculine supreme being. Along with that belief has come the suppression of women and the development of theologies declaring women to be naturally inferior."[109]

With this shift in spiritual orientation came a devaluing of the woman and the loss of social and political clout. Women, says Gimbutas, were for the first time subordinated within domestic hierarchies and made subject to strict sexual taboos. Often this denigration of women is connected to reproductive and ecological innovations, with clear parallels to contemporary issues. For example, it is surmised that, originally, Old Europeans did not understand the very notion of paternity and considered childbirth to be a miraculous accomplishment of the mother alone.[110] This helped to enforce matrilineal and matrilocal living patterns, and marginalize the control of men over the household—as, in the absence of paternity, it could scarcely be called "their" household.

But with the advent of animal husbandry, the sacred mystery of reproduction and birth were banalized, and placed under the control of instrumental reason. This paralleled another avenue for male supremacy, as increasingly intensive agriculture (using the plow and draught animals rather than the hoe) favored the superior upper body strength of the average male.[111] Thus, a complex of modern traumas (the loss of feminine sacrality and ecologically destructive farming, to name but two) were insinuated into the archeological narrative. Other anachronisms include Gimbutas's linking Neolithic and Bronze Age events to fifteenth-century European witch burnings, and even the brutality of Stalin's rule in Eastern Europe.[112] None of this is to downplay the contemporary and historical denigration of women. The only question, here, is whether Gimbutas's specific claims are, in fact, anachronistic.

These sudden, normative leaps into the present are endemic to Gimbutas's work. She will frequently draw associations between matrifocal religion in

---

109. Gloria Orenstein, David B. Axelrod, and Lenny Schneir, *Merlin Stone Remembered: Her Life and Works* (Woodbury, MN: Llewellyn, 2014), 100.
110. Marija Gimbutas, *The Living Goddesses*, ed. Miriam Robbins Dexter (Berkeley: University of California Press, 2001), 112.
111. Cynthia Eller, *The Myth of Matriarchal Prehistory: Why an Invented Past Won't Give Women a Future* (Boston: Beacon Press, 2006), 46–47.
112. Gimbutas, *The Language of the Goddess*, 318–19.

Neolithic Europe with modern European folktales and Baltic harvest rituals.[113] But, for Gimbutas and many of her followers, this is all part of a multi-millennia long, grand narrative.

In his preface to Gimbutas's *The Language of the Goddess*, the influential mythologist Joseph Campbell invokes the notion of the 5000 year horror of androcracy. Citing James Joyce, he bemoans "the 'nightmare' (of contending tribal and national interests) from which it is now certainly time for this planet to wake."[114] This is a common trope within the Goddess movement at large; "5000 years" of uninterrupted patriarchy allows feminist spiritualists to project contemporary political categories onto Neolithic Europe. After all, according to this view, politics is not timely, but universal in the most abstract, mystified sense. It is about the struggle between basic archetypes and energies, rather than the mundane conflict of material forces.

However well intentioned this narrative might be, it certainly has its critics—especially among certain feminists themselves. A particularly compelling and well-researched analysis of the Goddess movement is Cynthia Eller's *The Myth of Matriarchal Prehistory* (2000). While a thorough evaluation of Gimbutas's claims goes beyond the scope of this present book, it is worthwhile summarizing several of Eller's key criticisms. Eller's work underscores our central contention that Gaianism is, in truth, a modern affair. Neolithic matriarchy is less a discovery than a projection of modern wishes onto an imagined past. It may be true that the past 5000 years were a horror show of patriarchy; this does not imply, however, that prior to 5000 years ago there was an egalitarian utopia.

First, Eller points out that not every ambiguous figure discovered in "Old European" excavation sites is actually feminine. Gimbutas will claim that everything from bull figures, to depictions of horns, to abstract designs such as wavy lines and dots, and even sculptures of phalluses are representations of the Goddess Creatrix. (Phalluses are termed by her as "phallic goddesses," while bull horns are allegedly feminine because they vaguely resemble fallopian tubes.) She even describes figurines with no gender markers as definitively female, merely because they are not unambiguously male.[115]

Second, even in cases where female figures are found, there is not always clear evidence that these are sacred objects representing a goddess or female deity. As Eller points out, many of these objects fit the typical model of durable, children's toys, namely dolls. Alternatively, they may have represented nonsacralized mortal women, or even objects of sexual desire for men (a

---

113. Gimbutas, "The Earth Fertility of Old Europe," 16–20.
114. Gimbutas, *The Language of the Goddess*, xiv.
115. Eller, *The Myth of Matriarchal Prehistory*, 118–19, 128, 146–47.

hypothesis particularly odious to the Goddess movement).[116] This counter-hypothesis is plausible, as many of the excavated figures are not obviously pregnant (representing the "Creatrix" function), and some are in fact rather skinny.[117] Gimbutas, at times, clearly attempts to force the archeological findings to fit her predetermined narrative. For example, she will insist that a female figure with a large buttocks (not visibly pregnant) nonetheless represents the divine Creatrix because each cheek of the buttocks actually represents a sacred womb or egg. This doubling of fertility symbols supposedly represents the mystical "power of two" common to such societies.[118] Why this might be the case is never explained.

Third, even if there were depictions of female divinities associated with creation (a not-altogether implausible scenario), there is still no evidence that these all represent a single Goddess Creatrix. The notion of Goddess-monotheism is rarely supported by ethnographic evidence, with a few exceptions like the Hindu tradition of Shaktism. The majority of examples from antiquity, such as in ancient Greece, confute Gimbutas's theory; here we do see the depiction of female divinity, but no indication whatsoever of a singular Creatrix.[119]

Fourth, even if there was a matrifocal "Great Goddess" religion, that doesn't itself imply matrilineal living patterns or a matrilocal culture. In fact, matrilineality and matrifocality can easily be combined with other, highly patrifocal customs within the same culture, and vice versa. One needn't look further than matrilineal Judaism (religion is passed down from the mother), which is at the same time traditionally patriarchal and centered upon the worship of a male warrior-God. Similar combinations of male- and female-oriented customs abound worldwide:

> For example, the matrilineal Nairs "worship only male ancestors"; the patrilineal Mundurucú settle matrilocally, while the Trobrianders settle patrilocally; in Wogeo, New Guinea, potential marriage partners are selected matrilineally, but succession of political office and inheritance of property are patrilineal [...] Impressively, kinship can even be matrilineal in groups that insist that women are only passive carriers of men's seed, and patrilineal in groups that swear that men have no procreative role.[120]

---

116. Eller, *The Myth of Matriarchal Prehistory*, 133, 139.
117. Eller, *The Myth of Matriarchal Prehistory*, 134.
118. Gimbutas, "The Earth Fertility of Old Europe," 24.
119. Eller, *The Myth of Matriarchal Prehistory*, 103.
120. Eller, *The Myth of Matriarchal Prehistory*, 101–2.

Besides this, Eller notes that matrifocal religion may have little to do with women's own spiritual needs, but instead, represents men's sexual desire or the longing for a nurturing mother figure. Goddess religions, or female sacrality in general, may not indicate a heightened status for mortal women, but may only compensate for their low social status here on Earth. Eller cites, as an example, the reverence for the Virgin of Guadalupe in Mexico who is worshipped far more by men than by women.[121]

Fifth, even if there were matrilineal or matrifocal living patterns, that doesn't necessarily imply matriarchy. Where one lives or how family names are passed down, in other words, does not always correlate with actual power in society. Eller cites the matriliny imposed on slaves in the United States, which far from representing female power, was a lever of control over enslaved families (ensuring that the offspring of female slaves and slave masters belonged to the latter *as slaves*).[122] Moreover, merely owning and passing down land (in a matrilinear fashion) may not carry the same social weight as it does in modern society. In horticultural societies, land was less of an important store of wealth and prestige, as fields were often exhausted and households had to frequently move around.[123]

Sixth, even in cases of matriarchy (a female head of state, for example), there is no necessary correlation with gender equality for all classes of women. Clear examples of this include the highly inegalitarian societies of Victorian England and Thatcherite Great Britain. In any case, as Eller reports, the fact of matriarchy is unclear when it comes to "Old European" excavation sites. That women were found to be buried under large platforms with the rest of the family and men buried alone under smaller platforms *may* indicate (as Gimbutas suggests) that the family was the belonging of the female as head of household. Alternatively, it may equally suggest that all subordinate family members were crowded together in a common grave, with the dominant male having a reserved, private plot for himself.[124]

Seventh, even if there was gender equality in Old Europe, then this on its own says nothing about the absence of slavery, poverty, warfare, or general strife. Here, a Marxist analysis is useful. Supposing Neolithic Europe had greater gender equality than we see in modernity, it may not be a sign of an advanced civilization. To the contrary, pervasive scarcity may prohibit meaningful divisions of labor and the concomitant inequalities in wealth and social

---

121. Eller, *The Myth of Matriarchal Prehistory*, 104–6.
122. Eller, *The Myth of Matriarchal Prehistory*, 102.
123. Eller, *The Myth of Matriarchal Prehistory*, 110.
124. Eller, *The Myth of Matriarchal Prehistory*, 100.

status.[125] In other words, not all egalitarian societies are flourishing societies. Some may simply be too poor for stable hierarchies to arise. One should not confuse the "equality" of premodern starvation with the egalitarianism of a post-scarcity future.

In any case, there is ample archeological evidence of violent death among prehistoric populations. Arrowheads have been found lodged inside skulls in mass graves in Neolithic Europe, and there is possible evidence for human sacrifice in Minoan Crete (supposedly the paragon of peaceful, matriarchal society).[126] Gimbutas's claim that Neolithic Europe was devoid of defensive fortifications has been challenged by other archeologists. But even if this latter point were true, it would not preclude the possibility of offensive warfare waged on other people's territory. For example, Minoan Crete—an Island civilization—may have practiced naval warfare, leaving behind no land fortifications for archeologists to discover.[127]

Even Joseph Campbell, one of Gimbutas's most ardent advocates (and an enthusiast for the Goddess hypothesis) admits that a religion focused on the cyclical patterns of creation in Nature may be rather bloody.

> You regard the world of vegetation, and here are all these rotting leaves and rotting sticks in the jungle, and out of it come fresh shoots, and so the notion that out of death comes life leads to the next conclusion, that if you want to increase life, you need to increase death, and so there is an absolute frenzy of human sacrifice right across this glorious mother-right tradition. We always think of the Mother Goddess as so tender but just remember her with the axes in Crete, for example.[128]

Criticism of the Goddess hypothesis extends to scholarship of the ancient Near East. As we have seen, parallel to the theory that matriarchal Old Europe was displaced by brutish Kurgan pastoralists, there is a similar theory about the foundation of Western monotheism. We have here the image of a nature-worshipping, Canaanite matriarchy

---

125. This is analogous to Marx's critique of "crude" communism. Mere "levelling-down" is insufficient for the material or spiritual emancipation of humanity. Such fetishism of minimizing all wants and needs (common to certain utopian socialists) "not only failed to go beyond private property, but has not yet even to attain it." See Karl Marx, "Economic and Philosophical Manuscripts of 1844," in *The Marx-Engels Reader*, ed. Robert C. Tucker, Second Edition (New York: W. W. Norton, 1978), 83.
126. Eller, *The Myth of Matriarchal Prehistory*, 113.
127. Eller, *The Myth of Matriarchal Prehistory*, 114–15.
128. Joseph Campbell, *Goddesses: Mysteries of the Feminine Divine*, ed. Safron Rossi (Novato, CA: New World Library, 2013), 183.

which is conquered by the pastoralist Israelites with their warlike cult of Yahweh.[129]

It is true that Israelite religion underwent a reform movement after the period of Babylonian captivity (c. 537 CE). It is also true that this reform movement was hostile to the matrifocal practices of domestic piety common to the region.[130] Nonetheless, there is wide scholarly consensus today that nothing like the conquest of Canaan (as described, for example, in the Book of Joshua) ever occurred. The Israelites did not invade a harmonious, Gaia-centric culture from without. Ironically, the only groups which maintain this polemical myth tend to be feminist spiritualists on the one hand, and biblical fundamentalists on the other.

The real story is far more complicated—and interesting. Mainstream archeologists, such as William G. Dever, have pointed out that the Israelites were largely, themselves, of Canaanite origin. Furthermore, Canaanite spirituality certainly had its feminine aspects, represented especially in the figure of Asherah (Athirat).[131] But unlike the monotheistic Goddess hypothesis, Asherah was in fact the wife or consort of the prime male God, El. For his part, El was often associated with the sky, especially the sun, as well as the bull—often being referred to as "Bull El." So much for the bull figure having always represented femininity. And so pre-Israelite (i.e., Canaanite) society was itself rather patriarchal, further calling into question the Manichean narrative of modern Gaian scholarship.[132]

The Goddess hypothesis falters, in an even more striking way, when we closely examine religious traditions from the Indian subcontinent. Again, it is likely that Proto-Indo-Aryans did emigrate from somewhere around the Pontic steppe and into the Indus Valley. Furthermore, excavations of the Indus Valley Civilization (especially the cities of Mohenjo-Daro and Harappa) revealed an indigenous culture which was focused on the life-patterns of both flora and fauna. This civilization would likely give rise to the Dravidian cultures and languages of southern India and certainly predates the Vedic religions innovated by the arriving Indo-Aryans.

---

129. Merlin Stone, *When God Was a Woman: The Landmark Exploration of the Ancient Worship of the Great Goddess and the Eventual Suppression of Women's Rites* (New York: Harcourt, 1976), xvii–xx.
130. William G. Dever, *Did God Have a Wife? Archaeology and Folk Religion in Ancient Israel* (Grand Rapids, MI: William B. Eerdmans, 2005), 205–99.
131. Dever, *Did God Have a Wife?*, 100–102, 150, 166–67, 176 239.
132. Dever, *Did God Have a Wife?*, 101, 186, 210; William G. Dever, *Who Were the Early Israelites and Where Did They Come From?* (Grand Rapids, MI: William B. Eerdmans, 2003), 128; Frank Moore Cross, *Canaanite Myth and Hebrew Epic* (Cambridge, MA: Harvard University Press, 1997), 15, 32–34, 75.

Nevertheless, the findings from Indus Valley excavation sites do not confirm a Goddess religion. Images and figures found at Mohenjo-Daro and Harappa instead depict a male deity with a bull head (sometimes termed "Proto-Siva"), and female figures which were often skinny, wearing a "mini-skirt," and arranged in sensual positions—hardly the maternal Creatrix figure one would imagine, given Gimbutas's theory.[133]

In modern India, we still see matrifocal religious traditions, especially that of Shaktism. But the earliest textual sources for Shaktism are to be found in the Rig Veda, which of course was introduced by the "interloping" Indo-Aryan pastoralists. When we examine this text, it is highly reminiscent of ancient Greek mythology, especially as found in Hesiod himself. While the feminine is associated with primordial, oceanic Being, there is still a notable position for the male as lord of the sky and heavens above.

> I am the Queen, the gatherer-up of treasures, most thoughtful, first of those who merit worship […]
> They know it not, but yet they dwell beside me. Hear, one and all, the truth as I declare […]
> I make the man I love exceeding [sic] mighty, make him a sage, a Ṛishi, and a Brahman.
> I bend the bow for Rudra that his arrow may strike and slay the hater of devotion.
> I rouse and order battle for the people, and I have penetrated Earth and Heaven.
> On the world's summit I bring forth the Father: my home is in the waters, in the ocean.[134]

Though the feminine represents the entire godhead, she is nonetheless depicted as a helpmate to male figures which are placed literally above her. She is always "bending the bow for Rudra" (a male deity associated with storms), while offering enlightenment and sage-status to others.

When it comes to contemporary worship within Hinduism, there is likewise no necessary correlation between the worship of feminine figures, on the one

---

133. Sir Mortimer Wheeler, *The Indus Civilization*, Third Edition (Cambridge: Cambridge University Press, 1968), 86–90, 106; Gregory L. Possehl, *The Indus Civilization: A Contemporary Perspective* (Plymouth, UK: AltaMira Press, 2002), 141–44. The identification of this bull figure as a "proto-Siva" has long been controversial.
134. Ralph T. H. Griffith, trans., "Hymns of the Rig Veda," in *The Vedas* (New Delhi: Kshetra Books, 2017), Mandala 10, Hymn 125.

hand, and social esteem for actual women on the other.[135] Even Campbell, recalling his sojourn in India, reports:

> I learned that all women are divinities. The three great crimes in India are killing a cow, killing a brahmin, and killing a woman, because they all represent the sacred powers. Of course when you go to India you realize you can be very, very sacred and yet be in a rather inferior social position, but that's the incongruity of life—a mystery.[136]

But Campbell's shoulder-shrugging in the face of rank misogyny is insufficient. If one wants to promote a sweeping, trans-millennia, transcontinental theory of matriarchy, one must do better. The all-too neat narrative affirmed by the Goddess movement does not stand up to scrutiny—neither in "Old Europe," nor in pre-Vedic India, nor in pre-Israelite Canaan. Everywhere one looks, one sees goddess worship alongside powerful male deities; one sees patriarchy alongside cults of fertility; and one sees the pervasiveness of human violence and brutality alongside any number of spiritual customs.

As Eller relates a conversation with a Hindu pilgrim, "the difference between the goddess and women is like the difference between the stone you worship and the rock on which you defecate."[137] Clearly, religious practices rarely mirror everyday social values in a one-to-one correspondence. In any case, there can be no guarantee that ancient texts contain values appropriate to modern struggles for emancipation.

The problem with the Goddess movement is precisely in its "archeofuturism"—its faith that the past holds the key to the future. This is a term applied, almost exclusively, to the fringe Right-wing of politics. But the label fits here, if not in terms of policy positions, then in terms of basic methodology. Especially problematic is the persistent reification of gender differences and norms common to "Goddess" discourse. It is a repetition of the patriarchal idea that women are essentially peaceful nurturers, and that individual exploration and innovation are the purview of men. In other words, they reify women as "deep" and men as "powerful." This is hardly an advance over the conservatism of Hesiod himself.

---

135. Modern reactionary forms of political Hinduism, or "Hindutva," for example, affirm the worship of female deities while simultaneously promoting male supremacy in everyday social relations. Sikata Banerjee, *Make Me a Man!: Masculinity, Hinduism, and Nationalism in India*, SUNY Series in Religious Studies (Albany, NY: SUNY Press, 2005), 111–38.
136. Campbell, *Goddesses*, 16–17.
137. Eller, *The Myth of Matriarchal Prehistory*, 104.

And yet the Goddess narrative has persisted, especially outside of academia. One of the principal bridges between academic and popular Gaianism is Riane Eisler's book, *The Chalice and the Blade: Our History, Our Future* (1987). Following Gimbutas, *The Chalice and the Blade* repeats the story of a Bronze Age, matrifocal society grounded in Goddess worship predating the Kurgan invasions. This civilization is touted as having gender-egalitarian or "gylanic" features. And yet, the book sharply reinforces gender distinctions, where women are essentialized as having earthly roles as nurturing mothers. In this imagined world, the chalice is the main symbol of female power, representing the womb and the matrix of life. With the Indo-European influx, the female-centric power of the chalice is displaced by the blade, the symbol of male violence and domination.[138]

Perhaps unsurprisingly, Eisler's book was endorsed by Marija Gimbutas as "[a] notable application of science to the growth and survival of human understanding."[139] But *The Chalice and the Blade*'s impact was most widely evident in popular novels and films. In particular, Dan Brown's 2003 mystery novel, *The Da Vinci Code*, as well as its subsequent 2006 Hollywood film adaptation, make prominent use of Eisler's imagery.[140] Brown's story revolves around symbologist Robert Langdon in his quest to locate and safeguard the Holy Grail associated with the story of Jesus. Very quickly, however, it becomes evident that the "grail" in fact represents the sacred feminine, as against the inverted imagery of the blade—representing the phallic or the male.

> Langdon went on. "This icon is formally known as the *blade*, and it represents aggression and manhood. In fact, this exact phallus symbol is still used today on modern military uniforms to denote rank."
>
> "'Indeed.' Teabing grinned. "The more penises you have, the higher your rank. Boys will be boys."
>
> Langdon winced. "Moving on, the female symbol, as you might imagine, is the exact opposite.' He drew another symbol on the page. 'This is called the *chalice*.'

---

138. Riane Eisler, *The Chalice & the Blade: Our History, Our Future* (New York: Harper One, 1988), xvi–xix, 42–58.
139. Eisler, *The Chalice & the Blade*, i.
140. Dan Brown, *The Da Vinci Code* (New York: Knopf Doubleday, 2003); Katherine K. Young and Paul Nathanson, *Sanctifying Misandry: Goddess Ideology and the Fall of Man* (Montreal: McGill-Queen's University Press, 2010), 4.

[...] "The chalice," he said, "resembles a cup or vessel, and more important, it resembles the shape of a woman's womb. This symbol communicates femininity, womanhood, and fertility."[141]

Central to Brown's story is the idea that the grail or chalice is no mere artifact, but instead, represents the living bloodline of Jesus himself as carried on by his female descendants. This fact of female matriliny is suppressed, in Brown's novel, by an authoritarian and misogynistic Catholic Church. The novel and film adaptation were roundly denounced by many Christian denominations, and especially Catholic groups, as defamatory, just as it was celebrated by certain feminist thinkers at the time.[142]

This celebration of the sacred feminine was likewise extended into the realm of popular religion. This took many diverse forms. In some quarters, those who wished to remain within established, Judeo-Christian frameworks sought to discover or "read into" these traditions elements of feminine spirituality. In other cases, feminist spirituality involved a more radical break from, and criticism of, traditional religious forms and the founding of specifically matrifocal belief structures.

Examples of the former tendency include the works of Phyllis Trible, who recognizes the misogyny inherent in the Hebrew Bible, but nonetheless, believes that "the Bible can be redeemed from bondage to patriarchy; [and] that redemption is already at work in the text," especially in figures such as Eve and Miriam.[143] The Catholic feminist Rosemary Radford Ruether agrees that the Bible contains backward elements, but highlights what she calls "the covenantal and sacramental" traditions which are compatible with a feminist spirituality. Ruether reads the biblical account of Noah and the flood as a tale about humanity's relationship to Mother Earth. She also understands Jesus's statements about the body and the flesh as redignifying materiality.[144]

A more antagonistic tendency is to be found within 1970s-era "Dianic Wicca" as innovated by Zsuzsanna Budapest. This neo-pagan tradition distinguishes itself from other neo-pagans through worship of a unitary female

---

141. Brown, *The Da Vinci Code*, 237–38.
142. See, for example, the position of feminist biblical historian Elaine Pagels in "Opinion Page: Da Vinci Code' Truths," *Talk of the Nation* (Washington DC: NPR, May 22, 2006). Still other feminist thinkers criticized *The Da Vinci Code* for enforcing a stark gender dualism.
143. Phyllis Trible, "Eve and Miriam: From the Margins to the Center," in *Feminist Approaches to the Bible* (Washington DC: Biblical Archaeology Society, 1995), 8.
144. Rosemary Radford Ruether, *Gaia and God: An Ecofeminist Theology of Earth Healing* (New York: Harper Collins, 1992), 205–53.

Goddess based largely on the Roman deity Dianna, the goddess of the hunt and the moon.[145] It excludes males from the priesthood and generally eschews the worship of male pagan deities. Similarly, Carol Christ rejects Abrahamic religion as inherently patriarchal and oppressive. She concurs with Gimbutas that a peaceful, gynocentric religion once existed, and that it should exist once again in contemporary life.[146]

The impetus for a contemporary revival of matrifocal religion once again comes from speculative archeology. This is a tradition which predates even the efforts of Gimbutas. In the early twentieth century, Margaret Alice Murray advanced the so-called witch-cult hypothesis. Later discredited, the theory proposed that the European witch trials of Early Modernity were an attempt to stamp out a pervasive and "joyous" pagan religion for the benefit of an androcentric, repressed Christianity.

> Ritual Witchcraft—or, as I propose to call it, the Dianic cult—embraces the religious beliefs and ritual of the people known in late mediaeval times as "Witches" [...] It can be traced back to pre-Christian times, and appears to be the ancient religion of Western Europe. The god, anthropomorphic or theriomorphic, was worshipped in well-defined rites [...] it was a definite religion with beliefs, ritual, and organization as highly developed as that of any other cult in the world.[147]

Modern scholarship has never substantiated the existence of such a "well-defined" matrifocal religion, pervasive throughout all of Europe. However, this did little to impair the popularity of Murray's thesis. Nor does the dubious historicity of the witch-cult hypothesis dissuade many from using it as a basis for their present-day religious beliefs. While some adherents are unaware of the factual problems plaguing an ideologically driven archeology, others simply don't care. As Anne Carson opines:

> Let it be myth then [...] Whether the Golden Age of Matriarchy ever existed in history is not important: what is important is that the myth exists *now*, that there is a story being passed from woman to woman,

---

145. Zsuzsanna Budapest, *The Holy Book of Women's Mysteries* (San Francisco, CA: Red Wheel/Weiser, 2007), 62, 86–88, 246–60.
146. Carol Christ, *The Laughter of Aphrodite: Reflections on a Journey to the Goddess* (San Francisco, CA: Harper and Row, 1987), 117–31. For further discussion of this topic, see Ruether, *Gaia & God*, 149.
147. Margaret Alice Murray, *The Witch-Cult in Western Europe: A Study in Anthropology* (Oxford: Clarendon Press, 1921), 11–12.

from mother to daughter, of a time in which we were strong and free and could see ourselves in the Divine, when we lived in dignity and in peace.[148]

There is nothing wrong with taking myth *as* myth, that is, unlinking mythic narrative from the demand that it be literally true. Yet what counts, in the end, is the worldview that one's mythic narratives seek to illustrate. The narrative of a golden age of gender equality, dominated by feminine values, appears to be humane, especially as contrasted with the rank misogyny and sexual violence which plagues contemporary society. Yet on further consideration, such archeofuturism only reifies gender norms and thus a stark dualism between the masculine and feminine. It relegates female identity to a realm of misology, exoticizing the female as intuitive and deep, but nonetheless as the "other" of reason.

As for the actual existence of premodern gender parity, there is certainly evidence for this in many regions. At the very least, before the establishment of class society, there was often less of a stark distinction between the private household and the larger political economy. The communal household, therefore, was a center of production vital to the very reproduction of society itself. As organizers of the household, then, women automatically assumed a more public role which included all the customary dignities associated with it (religious, political, and otherwise).

But one should not romanticize this past which, after all, was still marked by material scarcity, warfare, and stringent communal limits on individual choice. "Primitive communism," as the Marxist anthropologist Eleanor Burke Leacock points out, is not to be facilely equated with utopia.[149] In any case, it could not last forever. Eventually, as the standard Marxist story goes, new and more productive technologies would threaten the stability of communal social relations—revealing the imperative for specialization in labor, and ultimately, the rise of class society. Barring that, it would only be a matter of time before "primitive communism" is conquered by other, more technically advanced civilizations (which, in fact, occurred numerous times).

For Friedrich Engels, the only way to achieve gender parity and personal freedom was by moving forward. Specifically, it will be the achievement of a post-scarcity society that will finally accord individual freedom to domestic relations and equality between men and women.[150] Only then will domestic

---

148. Anne Carson as cited in Eller, *The Myth of Matriarchal Prehistory*, 13.
149. See Eleanor Burke Leacock's Introduction to Frederick Engels, *The Origin of the Family, Private Property and the State*, trans. Alec West (New York: International, 2018), 25.
150. Engels, *The Origin of the Family, Private Property and the State*, 144.

ties be wholly unlinked to financial dependency. This would be a durable sort of freedom based in material abundance, rather than a fragile equality conditional on keeping society in a prolonged, artificial stasis. Too much of Gaian thought is merely the spiritualization of the latter sort of poverty, or what Marx called "barracks communism."

Scientific Gaianism

Alongside the Goddess movement's spiritualism was a scientific development of what would be called the "Gaia hypothesis." An important catalyst for this earth-centric hypothesis was, perhaps ironically, the US space program. The images which came back from the Apollo 8 mission of 1968, popularly termed "Earthrise," were particularly significant.[151] It took the space program, our literally leaving Earth, to inspire a renewed focus on our home planet.

Here, myth met cutting-edge science and technology. Joseph Campbell responded to these and similar NASA images with a sense of awe. Seeing our planet as a vital blue sphere within the cold expanse of space evoked a unitary feeling within him and many of his readers. With the backdrop of the Cold War, there was this shocking image of Earth as an internally integrated system—or perhaps even an organism itself—where all people are essential parts—not egoistic individuals or nations, but living in communion with one another, as well as animals, plants, and oceans.

> There you are, the whole planet as an organism [...] And the only myth that is going to be worth thinking about in the immediate future is one that is talking about the planet, not the city, not these people, but the planet, and everybody on it [...] And this would be the philosophy for the planet, not for this group, that group, or the other group. When you see the earth from the moon, you don't see any divisions there of nations or states. This might be the symbol, really, for the new mythology to come.[152]

But the term "Gaia hypothesis" did not have its origins in the comparative mythology of Campbell. Instead, the term was originally coined by then NASA scientist James Lovelock. In particular, Lovelock was tasked with determining which planets might show evidence of organic life. The

---

151. Robert Poole, *Earthrise: How Man First Saw the Earth* (New Haven, CT: Yale University Press, 2008).
152. Joseph Campbell and Bill Moyers, *The Power of Myth*, ed. Betty Sue Flowers (New York: Anchor Books, 1991), 40–41.

method he settled on was an analysis of the various atmospheres and their respective degrees of dynamism. Atmospheres which were largely static (such as those found on Venus or Mars) were deemed unlikely to support life, while dynamic atmospheres (such as Earth's) would be more promising.[153]

The major insight, then, was that both movement and self-regulation were key. In other words, "homeostasis" is the sine qua non of life. But if life is equivalent to homeostasis, then not only individual organisms but the whole of planet Earth must in some sense be "alive." This hypothesis of a self-regulating, living planet is what would become the so-called Gaia hypothesis.

As a professional scientist, Lovelock was always quick to qualify his more spiritually sounding remarks. For example, he denied the idea that Earth is a self-conscious or sentient being. At most, he affirmed a moderate teleology with regard to the "blue planet" insofar as its natural, internal processes are productive for the continuance of life.

> Occasionally it is difficult [...] to avoid talking of Gaia as if she were known to be sentient. This is meant no more seriously than is the appellation "she" when given to a ship by those who sail in her, as a recognition that even pieces of wood and metal when specifically designed and assembled may achieve a composite identity with its own characteristic signature, as distinct from being the mere sum of its parts.[154]

Humble though his rhetoric may sometimes seem, Lovelock has, over the years, advanced an ever more radical thesis. Originally, the "Gaia hypothesis" involved only the notion that life was productive for life. In other words, it claimed that organisms on Earth adapted to inorganic environments, and then affected those environments so that they would become gradually more hospitable for the growth of more diverse lifeforms.

Later, however, this notion was expanded into what Lovelock called the "Gaia *theory*." "Life does not merely regulate or make the Earth comfortable for itself." Instead, regulation "is a property of the whole evolving system of life, air, ocean, and rocks."[155] It is easy to miss the distinction here. What Lovelock is now suggesting is that the *entire planet*, both its organic and inorganic elements, are part of an integrated, teleological whole.

---

153. James Lovelock, *Gaia*, Second Edition (Oxford: Oxford University Press, 2016), 8–11, 36.
154. Lovelock, *Gaia*, xvi.
155. Lovelock, *Gaia*, 114.

One of Lovelock's most enthusiastic proponents, the philosopher Mary Midgley, illustrates this interconnection with reference to the carbon cycle:

> The carbon which living things use to form their bodies mostly comes, directly or indirectly, from carbon dioxide—the gas which, on the other planets, acts as a full-stop to atmospheric reactions. Life is therefore always withdrawing this gas from the atmosphere [...] if you stand on the cliffs of Dover, you have beneath you *hundreds of metres of chalk*—tiny shells left by the creatures of an ancient ocean. These shells are made of calcium carbonate, using carbon that mostly came from the air via the weathering of rocks."[156]

And so not only does organic life adapt to inorganic conditions, but the former is actually productive of the latter and vice versa.

> [O]n a planetary scale the coupling between life and its environment is so tight that the tautologous notion of "adaptation" is squeezed from existence. The evolution of the rocks and the air and the evolution of the biota are not to be separated.[157]

Here, though, one must take pause. For either the Gaian theorists are saying something empirically true, but philosophically trivial, or they are advancing a sweeping metaphysical claim without sufficient justification.

If, by the Gaia theory, Lovelock and Midgley only mean that living organisms both make use of, and sometimes revert to, inorganic matter, then this is a theory few scientifically literate people will take issue with. But if their theory instead claims that the "purpose" or telos of the environment as a whole is to be productive of life, then there just isn't evidence for such an outsized assertion. The latter, grander claim does confer a type of sentience upon the planet as a whole—however strenuously Lovelock and Midgley deny this. Barring this second sort of claim, their theories amount to very little beyond what can be found in a standard biology textbook. So either the assertion is that processes such as the carbon cycle exist (uncontroversial) or it is that planet Earth actively desires its own, fruitful existence (unproven). The latter, holistic picture is clearly in line with their overall rejection of a mechanistic view of nature. For Lovelock, Midgley, and many others in this Gaian community,

---

156. Mary Midgley, *Gaia: The Next Big Idea* (London: Demos, 2001), 16.
157. James Lovelock as cited in Midgley, *Gaia*, 37.

"holism" must overturn the mechanism of René Descartes, Galileo Galilei, and other Enlightenment thinkers.[158]

Gaian thought, likewise, overturns the basic schema of Darwinian evolution. For the Darwinian model is also highly mechanistic, founded upon a notion of efficient cause and effect. There are random mutations within organisms and those mutations which are best fit for their respective environment get passed along to the next generation at higher rates. This is wholly in accord with Enlightenment determinism. But this "blind" and unguided process is not the same as the seemingly self-aware or "cybernetic process" touted by Lovelock.[159] Lovelock uses the term cybernetic to describe distributed intelligence and intentionality throughout nature. Adaptation is therefore understood as purposeful—however cagey Lovelock is with his verbiage.

But intentionality and design are unnecessary hypotheses given Darwinian models of evolution. In fact, if one can speak of scientific theories as "having a point," then the point of the Darwinian model is to displace the earlier, theistic models of intelligent "design." And so, for all of the Gaian criticisms of Abrahamic and androcentric religion, Lovelock's cybernetic theory is strikingly reminiscent of the "Watchmaker arguments" of eighteenth-century Christian apologists.[160] He brings back the unnecessary and unprovable hypotheses of intention, design, and authorship with regard to the creation of life in the universe.

To the Gaian, however, the Darwinian model carries the risk of seeing nature as a plane of competition between hostile forces. Midgley, in particular, elides any difference between Darwinian thought to *Social* Darwinism. Hence, "survival of the fittest" implies an economic regime of laissez faire capitalism, the domination of men over women, and a brutal competition between nations. Supposedly, the "machine-imagery" at its basis is the impetus for feelings of "selfishness, spite, exploitation, manipulation," and other antisocial vices.[161] These retrograde attitudes can only be overcome with a countervailing image of the world as a cooperative, self-regulating whole. But, Midgley warns, the misogyny endemic to our present society will be hostile to such a move. This is particularly the case because the chosen image to represent this holism—Gaia herself—is female. As Midgley puts it, "does the scandal lie not so much in the subject-matter as in the sex of the deity? Is the idea of a female power in the cosmos somehow more unscientific than a male one?"[162]

---

158. Midgley, *Gaia*, 19.
159. Lovelock, *Gaia*, 45.
160. William Paley, *Natural Theology* (Indianapolis, IN: Bobbs-Merrill, 1963).
161. Midgley, *Gaia*, 14.
162. Midgley, *Gaia*, 23.

Such criticisms may be bewildering to her scientific and philosophical targets. For Midgley lumps together capitalists, communists, racial Darwinists, atheists, mechanists, and male-god-worshipping theists with little care for the major distinctions among them. She includes Marxists and socialists in this rogues' gallery as well—for they too are accused of affirming an intelligible, deterministic universe.[163] Such sweeping indictments, despite being clothed in scientific jargon, merely repeat the same mystical dualisms of Gimbutas and Eisler. The feminine, intuitive, bodily, cooperative, and holistic is counterposed to the male, analytic, intellectualist, competitive, and mechanistic. In the end, Midgley tips her hand when she admits that the Gaia hypothesis is as much a scientific concept as it is an inspiring, religious creed.[164]

In his 2000 preface to *Gaia*, Lovelock embraces what he calls a "postmodern" science before its "love affair with reductionism."[165] He approvingly cites the Czech playwright and statesman Václav Havel for recognizing Gaia as an alternative to rationalist ideologies (whether capitalist or Marxist). For Havel, rationalism and Marxism are deleterious to the human condition. As he puts it: "Our destiny is not dependent merely on what we do for ourselves but also on what we do for Gaia as a whole. If we endanger her, she will dispense with us in the interests of a higher value—life itself."[166] In acknowledging humanity as part of a greater superorganism, we can learn our proper place on Earth, lest Earth itself decides to eliminate us first.

But the trajectory of Gaian thought into the twenty-first century has revolved around this central tension: Gaianism seeks to promote a political program for humans, but is founded upon a rigorous anti-humanism.[167] This anthropological pessimism, however, is not the preserve of the Gaians alone. Anti-humanism, as we shall demonstrate, marks both the contemporary theories of eco-pessimism and accelerationism, both present-day Gaians and Prometheans. Each philosophical camp seeks to transcend human problems by transcending humanity itself.

---

163. Midgley, *Gaia*, 7, 9, 14, 26–27.
164. Midgley, *Gaia*, 23–24, 33–36.
165. Lovelock, *Gaia*, xviii.
166. Lovelock, *Gaia*, xv.
167. As for Havel, his own thought showed a tension between humanist and biocentrist tendencies. This is evidenced by the conflicting principles he espoused in his Liberty Medal acceptance speech (1994). Here, Havel affirms both the Gaia theory as well as the anthropic cosmological principle that humans "are not [merely] an accidental anomaly" in nature. See Lovelock, *Gaia*, xv–xvi.

Chapter 2

ACCELERATIONISM

Up to this point, we have been engaged in historical reconstruction. We have traced the evolutionary paths of the Promethean and Gaian myths, and their various mutations from antiquity to modern capitalism. Our task now is to critically evaluate these ideas—to move from history to philosophy. This will require a change in our method. The point of philosophy is to not only comprehend ideas, but to provide a normative analysis of the same. Put frankly, are these ideas any good? Do they track with reality? Do premises lead to their putative conclusions, and are the fundamental premises themselves true?

When dealing with contemporary Prometheanism, philosophical analysis comes up against special challenges. In the twenty-first century, the standard-bearers of Promethean mythos have been the so-called accelerationists. As alluded to earlier, accelerationism is a highly eclectic ideology which spans both the political Left and Right. Right-accelerationists, such as Nick Land, Mencius Moldbug, and Justin Murphy, hope to reach new horizons by exacerbating the processes of capitalism and even white nationalism.[1] Left-accelerationists, like Ray Brassier, Reza Negarestani, and the Laboria Cuboniks collective, by contrast, hope to accelerate past capitalism by democratizing productive technologies. Meanwhile, "Blaccelerationists" emphasize the historical exclusion of black people from white humanist discourses, and the historical process whereby capitalism has engendered the "black nonsubject." These are treated, not simply as moral wrongs, but rather as levers for a radical emancipation.[2] Whatever their political outlook, accelerationists tend

---

1. Nick Land, "The Dark Enlightenment," *The Dark Enlightenment* (blog), December 25, 2012, http://www.thedarkenlightenment.com/the-dark-enlightenment-by-nick-land/; Justin Murphy, *Based Deleuze: The Reactionary Leftism of Gilles Deleuze* (Other Life, 2019).
2. Aria Dean suggests that blaccelerationism "takes a long view of history wherein [...] living capital, speculative value, and accumulated time stored in the bodies of black already-inhuman (non)subjects" will be the agents of a future revolution. Critically, this is in contradistinction to locating the agent of revolution in some "inherited image of the human." See Aria Dean, "Notes on Blaccelerationism," *E-Flux* 87 (December 2017), https://www.e-flux.com/journal/87/169402/notes-on-blacceleration/; see

to reject a narrowly conceived local politics and affirm large-scale, technologically infused transformation.

But alongside this preoccupation with "the new," even verging on a sci-fi break with humanity itself, accelerationists tend to deploy their ideas in highly eccentric ways. The result is a plethora of neologisms, either original or often borrowed from continental philosophy: Terms such as "rationalist inhumanism," "teleoplexy," "autonomous bootstrapping," "abstractify," "abductive non-monotonicity," and confounding phrases like "telecommercialised nomadic multiplicity" abound.[3] Other times, accelerationists will use strikingly physicalist language to describe their abstract ideas. These have included, "critical torsion," "reformatting," "closed positive feedback loop," "short-circuiting," "machinic," "navigational," "mutational driver," as well as opaque expressions such as "discontinuous cut in the fabric of ontological synthesis," to say nothing of "trajectories of a vectorial [...] and not rotational or circulatory [...] sort."[4]

Certainly, neologisms can sometimes be useful when trying to express novel approaches to problems or when dealing with new phenomena. However, in the hands of the accelerationists, the proliferation of new terms seems to have become an end in itself. This has to do with accelerationism's cyberpunk origins in the Cybernetic Culture Research Unit (CCRU) at Warwick University in the mid-1990s.[5] If terms are hard to understand, or do not jibe

---

also McKenzie Wark's discussion of Kodwo Eshun's writings in *Sensoria: Thinkers for the Twenty-First Century* (London: Verso, 2020), 21–35.

3. Peter Wolfendale, "The Reformatting of Homo Sapiens," *Angelaki: Journal of the Theoretical Humanities* 24, no. 1 (February 2019): 58; Nick Land, "Teleoplexy: Notes on Acceleration," in *#ACCELERATE#: The Accelerationist Reader*, ed. Robin Mackay and Armen Avanessian (Falmouth, UK: Urbanomic Media, 2014), 509–20; Reza Negarestani, "The Labor of the Inhuman," in Mackay and Avanessian, *#ACCELERATE#*, 457; Patricia Reed, "Seven Prescriptions for Accelerationism," in Mackay and Avanessian, *#ACCELERATE#*, 535; Negarestani, "The Labor of the Inhuman," 437; CCRU, "Swarmachines," in Mackay and Avanessian, *#ACCELERATE#*, 326.

4. Ray Brassier, "Concrete-in-Thought, Concrete-in-Act: Marx, Materialism and the Exchange Abstraction," *Crisis & Critique* 5, no. 1 (2018): 115; Wolfendale, "The Reformatting of Homo Sapiens"; Negarestani, "The Labor of the Inhuman," 457; Ray Brassier, *Nihil Unbound: Enlightenment and Extinction* (Hampshire, UK; New York: Palgrave Macmillan, 2007), 139; CCRU, "Swarmachines," 330; Nick Srnicek and Alex Williams, "#Accelerate: Manifesto for an Accelerationist Politics," in *#Accelerate: The Accelerationist Reader*, ed. Robin Mackay and Armen Avanessian, Second Edition (Falmouth, UK: Urbanomic Media, 2017), 353; Laboria Cuboniks, *The Xenofeminist Manifesto: A Politics for Alienation* (London: Verso, 2018), 83; Brassier, *Nihil Unbound*, 149; Reed, "Seven Prescriptions for Accelerationism," 525.

5. Benjamin Noys, *Malign Velocities: Accelerationism and Capitalism* (Winchester, UK: Zero Books, 2014).

with present academic lexicons, then all the better. The ethos, here, is one of escaping from the "Cathedral" (or "Vampire Castle") that is professional academia and polite society.⁶ The counterculture medium is as much the message as is the content itself.

All of this makes it difficult to delineate what accelerationism is, and what its core arguments are—let alone to levy a critical analysis. Not only is the accelerationist crowd internally diverse, holding disparate political and theoretical commitments, but their use of signifiers is also not consistent between authors. In fact, there seems to be a contemporary embarrassment over the very term "accelerate" itself, with one accelerationist theorist calling it a "semantic disaster."⁷ One can never quite tell when idiosyncratic phrasing is deployed as a useful expedient, or alternatively, when novel phraseology is used as a road spike to troll the reader.

This is all to say that our method will be one of unapologetic reductionism. The term reductionism is perhaps one of the most repeated accusations in contemporary academia. It is invoked abstractly against authors who do not pay sufficient attention to the sublime complexity of the objects of their critique, whether this be real life or primary source texts. To be sure, it is possible to go too far with reductionism. Infamously, a crude "economism" reduces all social phenomena to technological production, as if ideas and social institutions had no causal power whatsoever.⁸ But while we should be on guard against oversimplification of this sort, we must reconcile ourselves to the fact that all understanding will involve some sense of reduction. To make anything intelligible, whether this be a natural occurrence or a theoretical text, there is a need to delineate what is central as opposed to what is peripheral. The imperative is to trace the actual connections between premises and conclusions, causes and effects, and not to stand in awe before celebrated texts and their authors.

If we are to make sense of accelerationist texts, we will need to deploy categories that, in many cases, are either ignored or rejected by the authors themselves. Given the inconsistent use of signifiers within the accelerationist milieu, it would be a fool's errand to try to tell a comprehensive story about

---

6. CCRU, "Cybernetic Culture," in Mackay and Avanessian, *#ACCELERATE#*, 317; Mark Fisher, "Exiting the Vampire Castle," *OpenDemocracy* (blog), November 24, 2013, https://www.opendemocracy.net/en/opendemocracyuk/exiting-vampire-castle/; Alex Williams, "Escape Velocities," *E-Flux Journal* 46 (June 2013), https://www.e-flux.com/journal/46/60063/escape-velocities/; Land, "The Dark Enlightenment."
7. Lucca Fraser, "XF Seminar Postmortem," *Feral Machines* (blog), May 3, 2020, https://feralmachin.es/posts/reflections_on_xf.md.
8. Richard K. Ashley, "Three Modes of Economism," *International Studies Quarterly* 27, no. 4 (December 1983): 463–96.

this ideological movement solely using its own vocabulary, for the vocabulary itself is fractured and forever shifting. A focus on signified over mere signifiers will be the only way forward.

To that end, the throughlines connecting accelerationist texts appear to consist of three principal interlocking categories. These are (1) Nominalism, (2) Voluntarism, and (3) Hyperstition. The remainder of this chapter will be an explanation of these three terms and a rigorous identification of how each function within various accelerationist texts. Finally, there will be a critical analysis of the social and political import of accepting this constellation of ideas. Namely, these concepts collectively tend toward an anti-humanism and an aestheticized politics at risk of turning reactionary.

As forewarned, our analysis will not likely be accepted by many of the accelerationist persuasion themselves. Objections will most readily come on the basis of vocabulary and word choice. For example, one of the leading accelerationist theorists, Ray Brassier, explicitly rejects the term "voluntarism."[9] Similarly, Peter Wolfendale dismisses the term "anti-humanism" in favor of his preferred label "inhumanism."[10] But in the spirit of putting ideas ahead of jargon, we will be translating accelerationist claims into a parlance more common to the wider philosophical community.

"Voluntarism," to take our first example, is commonly understood as the idea that the will can act without predetermined limits. This, as we will demonstrate, is something which Brassier affirms—even as he takes issue with the word "voluntarism" itself. Likewise, "anti-humanism" is commonly understood as the rejection of a stable and definable human essence. Wolfendale's preferred term, "inhumanism" does not escape this definition—whatever his protests may be.

Whenever possible, we will endeavor to highlight those times when our explanation of core accelerationist ideas elides the chosen signifiers of this or that author. For clarity's sake, this will often occur in the footnotes. For the main aim of this present work is to shine a light upon this seemingly obscure, but nonetheless influential, community of thinkers for the common reader and the nonspecialist.

Certainly, our analysis that accelerationist ideas tend toward an aestheticized politics, and do not sufficiently guard against political reaction, will be even more odious to those self-described *Left*-accelerationists. But that, after all, is the point of a critical analysis, that is, to discover the internal logic of an

---

9. Ray Brassier, "Prometheanism and Its Critics," in Mackay and Avanessian, *#ACCELERATE#*, 471.
10. Wolfendale, "The Reformatting of Homo Sapiens," 57; Negarestani, "The Labor of the Inhuman."

author's ideas despite that author's own subjective intentions. To paraphrase a Marxist slogan, just as "our opinion of an individual is not based on what he thinks of himself," so one cannot judge a theoretical work solely by an author's self-conception.[11]

Nominalism

Turning to our first concept, accelerationism is marked by a persistent nominalism. By this we understand the metaphysical stance that there are no objective and unchanging essences which define the stable contours of the universe. Whatever variations in vocabulary or theoretical nuances, accelerationists are united in dethroning a stable conception of Nature.

Among the Left-accelerationists, in particular, undermining the stability and unity of Nature has taken on a decidedly political tone. A prime example of this is the Xenofeminist tendency. In laudably defending the rights of queer and trans individuals (against both social conservatives and gender-essentialist feminists), the Xenofeminists seek to undermine what they perceive as the basis for essentialism itself: Nature.

> Nothing should be accepted as fixed, permanent, or 'given'—neither material conditions nor social discrimination forms […] Anyone who's been deemed "unnatural" in the face of reigning biological norms, anyone who's experienced injustices wrought in the name of natural order, will realize that the glorification of "nature" has nothing to offer us—the queer and trans among us, the differently-abled, as well as those who have suffered due to pregnancy or duties connected to child-rearing. XF [Xenofeminism] is vehemently anti-naturalist. Essentialist naturalism reeks of theology—the sooner it is exorcised, the better.[12]

These accelerationist-feminists criticize the gender/sex distinction of their mainstream counterparts. Whereas many feminists draw this distinction in order to oppose the alleged naturalness of traditional gender norms, and so reveal these as historically contingent, the Xenofeminists believe that this does not go far enough. In calling gender historically contingent, they surmise, one retains the "naturalness" and immutability of biological sex as its other. True emancipation consists in denying the natural givenness of biological sex as well. To do otherwise would be a false sort of liberation, one which enables a

---

11. Marx, "Preface to a Contribution to the Critique of Political Economy," 5.
12. Laboria Cuboniks, *The Xenofeminist Manifesto*, 15.

fluidity of gender roles, but which nonetheless disciplines the transgender individual. For they have not really escaped their nature-imposed sex identity, but *merely* their gender assignment. Thus, their slogan is "biology is not destiny."

> When the possibility of transition became real and known, the tomb under Nature's shrine cracked, and new histories—bristling with futures—escaped the old order of "sex." The disciplinary grid of gender is in no small part an attempt to mend that shattered foundation, and tame the lives that escaped it. The time has now come to tear down this shrine entirely, and not bow down before it in a piteous apology for what little autonomy has been won.[13]

Undoubtedly, the Xenofeminists are correct to stand up for the rights and welfare of transgender individuals. This is especially important given the rise of so-called trans-exclusionary radical feminists, colloquially known as "TERFS," in both academia and the popular press.[14] This exclusionary brand of feminism is characterized by a suspicion that transgender people are either insincere or deluded about their gender identities, and tends to de-emphasize the social harm which disproportionately affects them (including high-suicide rates, homelessness, and violence toward black transwomen in particular).[15] TERF ideology is founded upon a pseudoscientific gender-essentialism which reifies contemporary Western gender roles as permanent and normal.

While such unfounded gender-essentialism must certainly be criticized, it's not clear that this necessitates the abolition of all objective essences whatsoever. Such a sweeping theoretical move is not required to assert the historical fact that gender identities are fluid and change over time. Nature need not become unintelligible and anarchic in order to secure the rights of persecuted individuals. As we hope to establish in Chapter 4 of this book, a conception of humanism and universal human reason will be more productive for this sociopolitical goal.

Nonetheless, accelerationists of all stripes will take it as their basic position that there is no substantial reality to Nature. It is important, among this tendency, to counter the traditional Gaian position that Nature is a stable,

---

13. Laboria Cuboniks, *The Xenofeminist Manifesto*, 45.
14. Colleen Flaherty, "The Trans Divide," *Inside Higher Ed*, July 19, 2019, https://www.insidehighered.com/news/2019/07/19/divide-over-scholarly-debate-over-gender-identity-rages.
15. Serena Sonoma, "Black Trans Women Want the Media to Show Them Living, Not Just Dying," *Vox*, June 18, 2019, https://www.vox.com/first-person/2019/6/18/18679295/black-trans-women-murder-violence.

harmonious whole.[16] From the accelerationist perspective, "Mother Nature" is an oppressive fiction, one which hinders our capacity to take fate into our own hands. Stable conceptions of the universe, whether these be Newtonian or Gaian, are all part of a degraded theology of servitude. To escape this odious myth, one should pay more attention to how truly hostile and chaotic Nature can be. As opposed to the comforting ecocentric narrative, nature is "not a mother, but a stepmother who refuses to feed us."[17] It is only by virtue of our intellect and cunning that we survive on this planet at all.

> We live on a planet with two permanent polar ice caps, a planet whose land masses in large majority are stricken with snow, ice, freezing nights, and killing frosts every year, and whose oceans' average temperature is far below that of our life's blood. The Earth is a cold place. Our internal metabolism requires warmth. Yet we have no fur; we have no feathers; we have no blubber to insulate our bodies. Across most of this planet, unprotected life for any length of time is as impossible as it is on the moon. We survive here, and thrive here, solely by virtue of our technology.[18]

But such empirical insights, in the hands of the accelerationists, quickly turn metaphysical. Despite frequently degrading metaphysics as a holdover from theology, the accelerationist will often move from empirical premises to sweeping claims about the fundamental character of reality. While they sometimes claim to overcome metaphysics, they embrace an antinomian ontology which is every bit as speculative as what they reject.

A case in point is how a discussion of earth's hostile climate is transformed into an indictment of natural "Necessity" as such. In the words of Benedict Singleton, "Nature appears as the force of necessity" but one should not accept as necessary "*that which could be made otherwise.*"[19] In other words, not only can Nature be materially hostile to human beings, but it is the embodiment

---

16. It should be noted that the Earth as a harmonious whole is indicative of earlier Gaian thought, but tends to be rejected in favor of a more anarchic vision by contemporary Gaians. See Chapter 3 of this book.
17. Nikolai Federov, *What Was Man Created For? The Philosophy of the Common Task* (London: Honeyglen, 1990), 33; quoted in Benedict Singleton, "Maximum Jailbreak," in Mackay and Avanessian, *#ACCELERATE#*, 494.
18. Robert Zubrin, *Entering Space: Creating a Spacefaring Civilisation* (New York: Tarcher, 1999), 17–18; quoted in Singleton, "Maximum Jailbreak," 505–6.
19. Singleton, "Maximum Jailbreak," 495.

metaphysical determinism itself. Emancipation, therefore, consists in the radically free will surpassing all naturally imposed limitations.

The odd thing about this entire schema is that the accelerationist feels most threatened by what they deem to be the most unreal. Nature is at once an oppressive, murderous stepmother, and the enslaver of the human spirit, but at the same time, less than nothing—only a figment of the theological imagination. This tension typically comes down to whether the accelerationist author is speaking in social-political terms or scientistic ones. In the latter case, the "natural" is no longer seen as a threat, simply because the scientific disciplines nowhere find an object called "Nature." Yes, there are stars and rocks, lifeforms and bodies of water—but "Nature" does not exist.

> But science has no concept of "nature," and this is precisely what dissuades it from stipulating any limit between the natural and the extra-natural: nature is neither more nor less than the various discourses of physics, chemistry, biology, geology, ethology, cosmology [...] The list remains necessarily open-ended. Where the sciences of nature are concerned, the irreconcilable is their highest concept and the irremediable their only meaning.[20]

It is easy to read this stance as a scientific rejection of metaphysics. But nothing could be further from the truth. The proposition that the various scientific disciplines may be "irreconcilable," and so make the concept of Nature "irremediable," is itself metaphysically significant. It goes beyond an empirical horror at Nature's violence and goes to a direct criticism of the very notion of "substance."

In the context of the Enlightenment rationalism of Spinoza, substance denotes a being which is infinite and self-causing.[21] It is this kind of permanence and absolute intelligibility that the accelerationist opposes. In his *Ethics*, Spinoza's explicitly identified substance with infinite, lawful Nature. Natural laws are immutable precisely because they are universal. This is opposed to traditional theistic models wherein God's intervention in history is

---

20. Brassier, *Nihil Unbound*, 40.
21. Benedictus de Spinoza, "Ethics," in *The Collected Works of Spinoza*, Volume I, ed. and trans. Edwin Curley (Princeton, NJ: Princeton University Press, 1985), E1D3; references to Baruch Spinoza's *Ethics* will be according to his own designations in that work. "E" refers to the Ethics itself, followed by the book number. "D" refers to definitions; "A" to axioms, and "P" to propositions. Likewise, "dem." will signify a demonstration and "schol." a scholium.

capricious—a matter of His will—and where divine miracles can supersede ordinary causes and effects.[22]

But in opposing the existence of Substance, the accelerationist also does away with Substance's lawful necessity—that is, the principle of sufficient reason (PSR). Running the gamut from Nick Land's early work *Thirst for Annihilation*, toward later Left-accelerationist texts, we see a unified abhorrence of determinism, and the PSR in particular.[23] In Brassier's words, "intelligence culminates with the delirium of the sufficiency of reason."[24]

Nonetheless, the main achievement of early modern rationalism was a distinction between *Natura naturans* and *Natura naturata*, roughly translated as "naturing-nature" and "natured-nature." Arcane as this division may seem, it is nothing more than the distinction between, on the one hand, the laws of the universe, and on the other hand, the existing features of the world which are conditioned by those laws. Crucially, in this rationalist schema, *Natura naturans* (natural laws) are unchanging and sovereign relative to those particular events which they condition. Transient events can change, but the law does not. Nothing, as they say, is above the Law.

While some accelerationists may fetishize the signifiers of the Enlightenment, and of rationalism, they are nearly all united in rejecting the above position. Their own metaphysics, to the contrary, is one of denying "Substance" (with a capital "S") and natural law. Specifically, they tend to blur key distinctions within philosophy: between form and matter, being and becoming, cause and effect, actuality and potentiality, necessity and contingency, and between the possible and the impossible. Thus, a political critique of "Nature" reveals itself to be a deeper abhorrence of philosophical Substance as that which retains its identity beyond transient changes.

> [T]he self-estrangement of essence deformalizes substance [...] The subordination of becoming to substantial form is undone and the possible and impossible are desegregated. Only what has become can be retrospectively considered essential. And what has become essential retroactively determines what will be possible. Every becoming re-establishes

---

22. Spinoza, "Ethics," E1 Appendix.
23. Nick Land, *Fanged Noumena: Collected Writings 1987–2007*, ed. Ray Brassier and Robin Mackay, Second Edition (Falmouth, UK: Urbanomic, 2012), 159; Nick Land, *The Thirst for Annihilation: Georges Bataille and Virulent Nihilism* (London: Routledge, 2002), 6. 101; Brassier, *Nihil Unbound*, 64–66; Luciana Parisi, "Automated Architecture: Speculative Reason in the Age of the Algorithm," in Mackay and Avanessian, #*ACCELERATE*#, 419.
24. Ray Brassier, "Liquidate Man Once and for All," *In/Appearance* (blog), 2005, https://inappearance.wordpress.com/2009/11/03/liquidate-man-once-and-for-all/.

the limit between the possible and the impossible as a division set to be undone by the practical actualization of the essential difference that underlies it.[25]

What may at first appear to be a sophisticated, "dialectical" approach to philosophy turns out to be nothing of the kind. Simply conflating cause and effect, or the possible and the actual, in no way establishes the dialectical unity of these concepts. A rationalist dialectics would seek to explain how provisionally opposed concepts can be reconciled in a higher order of intelligibility. But the accelerationist agenda, given its nominalism, is constitutionally opposed to any such reconciliation or final logic. It therefore wears the clothes of dialectics (when fashionable), but has the beating heart of scholastic nominalism.

The outward appearance of Hegelianism among some accelerationists is something of a transformation from their original position. Early works such as Ray Brassier's *Nihil Unbound* (2007) and Reza Negarestani's *Cyclonopedia* (2008) stuck close to the techno-horror pessimism of the original CCRU project. The trend since then, in works such as "Strange Sameness" (2019) and *Intelligence and Spirit* (2018), has shown greater openness to Hegelian and even Marxist thought (all through the lens of Robert Brandom and the Pittsburgh school of Hegelianism). Despite this apparent transformation, the antinomianism and pessimism of their original projects remain largely intact.

Brassier's early statements, such that we "must be prepared to take the plunge into the black hole of subtraction," merely find a rhetorical updating in his newer works.[26] Now invoking "Enlightenment Prometheanism," Brassier locates its "cardinal epistemic virtue […] in recognising the disequilibrium which time introduces into knowing." The Enlightenment is praised for its "catastrophic logic" and dialectics is valued for its affirmation of "indissociability."[27] All seemingly dialectical resolutions will be provisional and open to further dissolution. Nature, for the accelerationist, is in no wise a coherent whole or "given."

In this, Left-accelerationist thought truly begins with Wilfrid Sellars. It is Sellars's rejection of the "myth of the given"—that our sense-perceptions can grant us reliable knowledge about the world at large—that informs their point of departure. If phenomenological experiences are dubious, then one must be suspicious of all manner of folk psychology and naive realism. As such, there is an opposition between the so-called manifest image and the scientific

---

25. Ray Brassier, "Strange Sameness: Hegel, Marx and the Logic of Estrangement," *Angelaki: Journal of the Theoretical Humanities* 24, no. 1 (February 1, 2019): 102.
26. Brassier, *Nihil Unbound*, 100–101.
27. Brassier, "Prometheanism and Its Critics," 469–70.

image.[28] The former denotes the accumulation of "common sense" notions and received wisdom (whether this be theological, metaphysical, or philosophical). The latter denotes only those items of knowledge which are derived from the hard sciences themselves. A classic example of this division of knowledge would be the manifest image of our subjective self-understanding (emotions, deliberative choice, etc.) as opposed to the scientific insights of neurobiology (brain states, the firing of neurons, and so forth).[29]

Certainly, it makes sense on some level to distinguish between the lived experiences of ordinary people and the verified, repeatable discoveries of the scientific method. There is no basis for claiming equality between the scientific observation of the Earth as spherical, as opposed to the everyday experience of one's immediate terrain as being flat.

That being said, the accelerationist division between the manifest and the scientific image is based on a more extreme assertion that there is no permanent, intelligible world that can be ultimately known. For them, rejecting "givenness" is not only to indict a naive empiricism. It is the further claim that the world beyond our senses is anarchic and irremediably alien; we are left with no comforting stability. "Like every myth of the given, a stable foundation is fabulated for a real world of chaos, violence, and doubt."[30] Under a naive realism, they assert, "The 'given' is sequestered into the private realm as a certainty, whilst retreating on fronts of public consequences."[31] In other words, we pretend that the world is not weird in order to maintain our parochial prejudices in everyday life.

Givenness is rejected, on theoretical grounds, because of the more basic claim that all knowledge is mediated. Here, the particular target is the philosophical tradition of phenomenology. Specifically, they accuse this tradition of ignoring how our everyday experiences are always, already theoretical rather than straightforwardly "given" to the senses. "Phenomenology's absolutization of given-ness as such is the most extreme variant of the myth dismantled by Sellars."[32]

Simply put, starting with sense experience is a doomed approach, according to the accelerationist. For it unjustifiably stacks the deck in favor of affirming one's own, preconceived notions. Besides, it ignores how even the manifest image itself is a "subtle theoretical construct."[33]

---

28. Wilfrid Sellars, *Science, Perception and Reality* (Atascadero, CA: Ridgeview, 1963), 1–40.
29. Brassier, *Nihil Unbound*, 12.
30. Laboria Cuboniks, *The Xenofeminist Manifesto*, 15.
31. Laboria Cuboniks, *The Xenofeminist Manifesto*, 45.
32. Ray Brassier, "The View from Nowhere," *Identities: Journal for Politics, Gender and Culture* 8, no. 2 (Summer 2001): 22.
33. Brassier, *Nihil Unbound*, 3.

The critique of phenomenology typically extends to Heidegger, whom they rightly characterize as placing the human frame before supposedly objective scientific rigors. The most famous example of this, of course, is Heidegger's rejection of the primacy of clock-time in favor of the experience of time unique to our being-in-the-world.[34] The accelerationist ridicules such subordination of technology to the human frame and human activities. At one point, Brassier sarcastically demands that the Heideggerian "explain precisely how, for example, quantum mechanics is a function of our ability to wield hammers."[35]

But for all that, the accelerationist ends up in a surprisingly similar position as their phenomenologically minded opponents. For given their basic nominalism, objective scientific knowledge has no priority over our actions in the world, including our discovery of new technology. The will is not restricted by the objective laws of an intelligible universe; instead, our free creations help to produce the world.

This Promethean attitude explains why the accelerationist affirms an ideology of technoscience. As the #ACCELERATE Manifesto puts it, "we surely do not yet know what a modern technosocial body can do."[36] The Xenofeminists likewise affirm that "nothing is transcendent or protected from the will to know, to tinker and to hack," and so technoscience serves as a further means to destroy all established hierarchies.[37]

But if Heidegger's hammer is derided as a parochial extension of the human frame, then it is unclear how the most advanced gadget touted by technoscience is really anything different. In both cases, a created object is said to have priority over objective and universal knowledge of the world. This existentialist conclusion is unavoidable given the accelerationist's own premises. Barring an orderly, intelligible universe, we are left merely with the rigors of "science." But what guarantees the scientific method? Not Nature or the world, but the scientific disciplines themselves. This vicious circularity lends itself to a valorization of scientific praxis for its own sake. And what is scientific praxis but the creation of ever new ways to manipulate and control the world, namely technology? Knowing has no priority over doing.

In a parallel way, the natural has no priority over the synthetic. The natural light of reason is replaced by the inorganic intelligence of advanced technology:

---

34. Martin Heidegger, *Being and Time*, trans. John Macquarrie and Edward Robinson (New York: Harper & Row, n.d.), 468–69.
35. Brassier, *Nihil Unbound*, 8.
36. Srnicek and Williams, "#Accelerate: Manifesto," 355–56.
37. Laboria Cuboniks, *The Xenofeminist Manifesto*, 65.

Thus the artificialization of intelligence, the conversion of organic ends into technical means and vice versa, heralds the veritable realization of second nature [...] the irremediable form wherein purposeless intelligence supplants all reasonable ends.[38]

Here, the antinomianism of A.I. exceeds Heidegger's parochialism. A software's intelligence becomes detached from human interests and welfare. But this can be read, not as a rejection of Heidegger, but rather as an "acceleration" of his own existentialism. If being human is just the project of a creative will, if a thick conception of human nature is discarded, then creativity itself is pre-eminent. There can be no a priori limits to what the products of this creativity can do. The machines don't have to be our friends. Thus, while accelerationism moves beyond Heidegger's Black Forest traditionalism, it fully embraces his critique of "onto-theology," that is, his prioritizing of pure possibility over actuality.[39]

This accelerationist emphasis on the possible over the actual frames their view on technology and its role in civilization. Technology is seen as liberatory because it breaks free from the rigors of formal logic and reason. At most, reason is "not a universal that can be imposed from above, but built from the bottom up—or, better, laterally, opening new lines of transit across an uneven landscape."[40]

Luciana Parisi writes of design technology as having the power to disclose new "spatio-temporal patterns" which go beyond a preconceived Reason. These are genuine novelties which supposedly exceed the technology's original programming. In order to affirm such radical newness, there is a prioritizing of materiality over form. The plurality of materials each possess their own unique potential, and these are not governed by any mechanistic laws. Instead, materials inform reason as its own emergent property.

As part of the generic tendency to accelerate automation, the turn to inductive reasoning [...] allows matter to become the motor of truth, to become one with and ultimately constitutive of formal reason, of the rules and the patterns that emerge in the automation of space and time.[41]

---

38. Brassier, *Nihil Unbound*, 47.
39. Virgilio A. Rivas, "Dasein, Objects, and Myth: Heidegger's Unconscious Accelerationism," *3:AM Magazine*, October 10, 2015, https://www.3ammagazine.com/3am/dasein-objects-and-myth-heideggers-unconscious-accelerationism/.
40. Laboria Cuboniks, *The Xenofeminist Manifesto*, 57.
41. Parisi, "Automated Architecture," 406.

But if matter (or "materials") is prior to form, then it is unclear what place reason has at all in such a system. One may insist on using the signifier "reason," but unless reason is that which describes the relations between particular materials in a universal way, and thus makes all material things comprehensible relative to one another, then the term lacks genuine meaning.

The Left-accelerationist's Promethean impulse is to break free from a dusty logic and a suffocating rationalism in order to radically transform the world, that is, to "*storm the heavens* and *conquer death*."[42] And yet, they deprive themselves of exactly that faculty—reason—which would be necessary to intervene in current politics and to determine our collective future. Within their schema, matter is necessarily aleatory and the future an "indeterminate entity." It appears to be their view that such indeterminacy is a feature and not a bug; incompleteness and uncertainty enable a state of play and the chance for heroic intervention.

> To speculate is to articulate and enable the contingencies of the given armed only with the certainty that what is, is always incomplete; to speculate is to play with the demonstration of this innately porous, nontotalisable set of givens.[43]

Nonetheless, one may ask how intervention is even possible in a world turned to quicksand. Not only is a consistent description of the state of things rendered impossible, but a normative framework appears equally elusive.

This is because a nominalism about the universe will naturally extend to a nominalism about human nature, and consequently, any stable notion of human welfare.

One of the most explicit treatments of accelerationist anti-humanism is provided within Reza Negarestani's "The Labor of the Inhuman" (2014).[44] In it, Negarestani rejects the entire concept of a human essence, or of any "predetermined idea of man."[45] Instead, the labor of the inhuman consists precisely in separating the "significance of the human" from any meaning "established by theology."[46] In this, Negarestani relies on the old device of

---

42. Singleton, "Maximum Jailbreak," 494.
43. Reed, "Seven Prescriptions for Accelerationism," 527.
44. Negarestani, "The Labor of the Inhuman." Negarestani prefers the term "inhuman" to "anti-humanist." However, as with Wolfendale, this is not because he rejects the common understanding of anti-humanism, but rather because he sees the term "inhuman" as more rigorously departing from humanist prejudices.
45. Negarestani, "The Labor of the Inhuman," 430.
46. Negarestani, , "The Labor of the Inhuman," 427.

equating any genuine rationalist or a priori philosophy with a mystical theism.[47]

Negarestani's antidote to humanism is, characteristically, existentialist. Invoking Jean-Paul Sartre, he assigns the human a "universal" quality only insofar as it freely defines itself through the projects it undertakes.[48] Negarestani likewise (in his later works) affirms the idea that there are certain inviolable norms of thinking. Even here, however, these norms aren't grounded in any objective metaphysics and are more like formal commitments than objective realities.[49] Ultimately, it is not we humans who undertake projects, but projects which undertake us: We are nothing apart from our commitments and what we choose to do.

Given the current tendency of accelerationism to adopt a Hegelian/Marxist veneer, this ethos of the "free, creative project" will typically be associated with revolutionary jargon. In Brassier's view, Marxian Prometheanism consists in "the project of re-engineering ourselves and our world on a more rational basis."[50] But again, it is not clear whether "rational" is here meant in a descriptive or normative way. Whatever the case, a universe marked by disequilibrium, pure potentiality, and indeterminate matter, cannot sustain either sense of the term "rational." What is and what should be are converted into purely free decisions, lacking any stable basis for criticism, approbation, or improvement.

Voluntarism

This discussion leads us to accelerationism's next key concept: voluntarism. Voluntarism is the idea that the will has priority over the intellect. We are radically free to act in the world without any limitations whatsoever, whether these take the form of natural laws or of the nature of our own minds (e.g., brain states). Of course, voluntarism is a necessary outgrowth of accelerationism's basic assertion of nominalism. If there are no universal essences, and the

---

47. Negarestani does affirm the signifier of "rationalism," but as per our above discussion, he rejects everything this term commonly signifies. In particular, he rejects the core notions of "foundationalism" and "metaphysical determinism" so central to Enlightenment rationalism. See Negarestani, , "The Labor of the Inhuman," 454.
48. Negarestani, , "The Labor of the Inhuman," 428.
49. Reza Negarestani, *Intelligence and Spirit* (Falmouth, UK: Urbanomic, 2018). For a capable discussion of Negarestani's evolution as a thinker, see Vincent Le, "Spirit in the Crypt: Negarestani vs Land," *Cosmos and History: The Journal of Natural and Social Philosophy* 15, no. 1 (2019): 535–63.
50. Brassier, "Prometheanism and Its Critics," 487.

world is irreducibly dynamic to the point of chaos, then one requires a radically free (literally supernatural) will to cohere this centrifugal picture.

Undoubtedly, accelerationists will deny that their philosophy includes a theistic will (or a religious notion of *creatio ex nihilo*). But the emergence of such a radically free will within a nominalist framework is, ironically, a conceptual necessity, and not a free choice.

Alternatively, some accelerationists may insist that their worldview is not one of *irreducible* dynamism, but rather contains local regions of consistency.[51] By this, they may hope to retain a degree of intelligibility against a nominalist backdrop. Nonetheless, on a conceptual level, this strategy is not promising. It does not solve the problem of intelligibility but merely postpones it to a second-order question. For by the accelerationist's own lights, there can be no stable, intelligible way to determine *which* local regions of the universe are intelligible and which are not, let alone why. To posit local islands of intelligibility within a sea of indeterminacy is itself unintelligible. For what accounts for these local regions of stability and coherence in the first place?

To belabor this analogy a bit: Where are these intelligible islands located? What is their distance or ratio to one another? How big are they? Can one accurately describe their borders? To think that one can answer such questions is to covertly undermine and tame the supposed dynamism inherent to accelerationist thinking. But to *retain* this dynamism, all "islands of intelligible stability" must necessarily disappear under the chaotic waves. Finally, the basis for stability will have to be something like the free will, that is, a force which creates out of nothing and is not itself accountable to anything.

Still, the more sophisticated of the accelerationists will sometimes eschew the term voluntarism altogether. Ray Brassier, for example, criticizes voluntarism not because he rejects the autonomy of the will, but rather because he wants to go even further in separating such autonomy from the human frame.[52]

In a parallel move, Reza Negarestani insists that reason is not wholly separate from nature. However, he quickly modifies this statement by claiming that "reason has irreducible needs of its own" and that "it can be assessed only by itself," (following both Kant and Sellars). Ultimately, reason is "autonomous" in the sense that its "first task" is to create for itself a conception of nature.[53] Hence, Negarestani engages in an extreme formalism. He may admit that rational beings once emerged from the primordial sludge of the natural

---

51. Brassier, *Nihil Unbound*, 79; Parisi, "Automated Architecture:," 405; Reed, "Seven Prescriptions for Accelerationism," 534.
52. Brassier, "Prometheanism and Its Critics," 471.
53. Negarestani, "The Labor of the Inhuman," 454.

world; however, once reason exists, it is entirely free to posit its own normative laws and even descriptive reality.

This is not the naturalism of Spinoza, which has reason mirroring the objective "order and connection of things." It is instead the transformation of reason as something sublimely independent of materiality, and so, literally "super-natural."[54] Whereas Spinoza's "God" or objective Nature is the measuring stick whereby we judge the adequacy of the mind's ideas, for Negarestani "the Good begins with the death of God [...] [the] cancellation of all given totalities in history."[55]

The Left-accelerationist has a very specific problem: On the one hand, they want to develop and harness new technologies in order to act in the world, while on the other, their stated goal is a radical emancipation which breaks through all natural limits. But while these two goals at first appear to dovetail with one another, they actually produce an irresolvable tension. Acting in the world in order to modify one's circumstances presumes an understanding of what the world "is." But their normative project is based principally in transcending all descriptive facts whatsoever. It's as if they throw a room into total darkness in order to better navigate it.

One strategy of avoiding this problem, among the accelerationists, is to draw a distinction between mundane and supermundane faculties of knowledge. This is a move, imitating Kant and Hegel, which posits a distinction between the work of the (mundane) Understanding and the activities of (supermundane) Reason. In Brassier's words, "it is the understanding, the faculty that dismembers, objectifies and discriminates, which [...] will be subsequently consummated by reason."[56] Put otherwise, the understanding registers and analyzes what "is," while reason applies our free commitments to the world, thereby transforming reality itself. In this way, the accelerationist once again hopes to have their cake and eat it too; one can both appreciate the "hard facts" but also radically transcend them.

When accelerationists affirm the signifier "reason," it is in this latter sense of transcending descriptive reality. We can abstract from our current frames in order to freely create the world we desire. In the hands of many Left-accelerationists, like Patricia Reed, this has definite political implications.

> First and foremost, abstraction is a separation from what *is* towards what *could be*. In this regard, it is a gesture of violence, an affirmative violence in exiting the as-it-is condition and moving towards the generation

---

54. Spinoza, "Ethics," E2P7.
55. Negarestani, *Intelligence and Spirit*, 505.
56. Brassier, "Prometheanism and Its Critics," 470.

of new connections to and with a world. The power of abstraction to experiment and revise relations to each other, to production, to value creation and to the world, is a capacity that needs to be reclaimed beyond its colonization by finance capital and labour relations.[57]

The point is to break free from all mundane strictures of "the given." Genuine scientific thinking, for them, no longer takes seriously Laplace's "clockwork universe," or the notion that we can master Nature merely by accumulating enough objective information.[58] Instead, as the #ACCELERATE Manifesto puts it, "The future needs to be constructed."[59] Progress is no longer seen as following a predetermined path, but rather, consists precisely in innovating ever new trajectories into an irreducibly open future.[60] Such voluntarism explicitly defines freedom as that which liberates itself from the "muck of immediacy."[61]

Freedom, understood in this way, is a "political achievement," but one with very definite metaphysical implications.[62] For it presumes not only a universe which is indeterminate and open to possibilities, but also that it is our task to make it so. It is by the force of the human will, imagination, and cunning that we throw off both fate and determinacy, and impose freedom upon the world. "The future must be cracked open once again, unfastening our horizons towards the universal possibilities of the Outside."[63]

Reason is often cast as that supermundane faculty which can affect such a rupture. In this way, the accelerationists definitively depart from the tradition of the Enlightenment, especially that of the most consistent Enlightenment rationalist Baruch Spinoza. Famously, Spinoza rejected voluntarism, or the idea that the will could be independent from the intellect.[64] In the *Ethics*, Spinoza writes, "The Mind is a certain and determinate mode of thinking, and so cannot be a free cause of its own actions, *or* cannot have an absolute faculty of willing and not willing."[65] Such rationalism demands that the will merely follow the dictates of the intellect, and cannot freely act on its own.

---

57. Reed, "Seven Prescriptions for Accelerationism," 536.
58. Srnicek and Williams, "#Accelerate: Manifesto," 360–61.
59. Srnicek and Williams, "#Accelerate: Manifesto," 362.
60. Nick Srnicek and Alex Williams, *Inventing the Future: Postcapitalism and a World Without Work* (London: Verso, 2015), 75.
61. Laboria Cuboniks, *The Xenofeminist Manifesto*, 15.
62. Srnicek and Williams, *Inventing the Future*, 75.
63. Srnicek and Williams, "#Accelerate: Manifesto," 362.
64. Spinoza, "Ethics," E2P48; E2P49.
65. Spinoza, "Ethics," E2P48 Dem.

While the accelerationists do not explicitly separate the will from the intellect, they functionally achieve the same effect. Their modus operandi is to fold reason, the will, and even the imagination into one faculty—but in doing so, they remove any priority of intellection over willing. In this way, all knowing is irreducibly a free choice, or commitment, or creation. The will becomes "free" after all. Whatever their Enlightenment pretenses, then, the accelerationists achieve exactly what Spinoza most strenuously warned against: a power of choosing which exceeds objective knowledge.

Since this is the case, any description of the world will likewise be irreducibly normative. Knowledge of the world, that is, how things *are*, will have no logical priority over how they *ought* to be. For intellection is entirely inseparable from our free, moral commitments.

> The distinction between the ontological and the normative, between fact and value, is [not] simply cut and dried [...] The project of untangling what ought to be from what is, of dissociating freedom from fact, will from knowledge, is, indeed, an infinite task.[66]

The necessity of reason, therefore, does not consist in its mirroring the objective laws of the universe. To the contrary, the necessity and autonomy of reason is to be understood only in the normative sense of freely choosing what ought to be.

> But what exactly is the functional autonomy of reason? It is the expression of the self-actualizing propensity of reason—a scenario wherein reason liberates its own spaces despite what naturally appears to be necessary or happens to be the case. Here "necessary" refers to an alleged natural necessity, and is to be distinguished from normative necessity. Whereas the given status of natural causes is defined by "is" [...] the normative of the rational is defined by "ought to be."[67]

Voluntarism means taking an "interventive attitude" toward norms that renders them binding.[68] This strongly echoes Kant's moral

---

66. Laboria Cuboniks, *The Xenofeminist Manifesto*, 67. The accelerationist may balk at the suggestion that they are conflating fact and value. Instead, as here, they may wish to appear more nuanced and only claim that distinguishing the two is an "infinite task." However, such semantic hair-splitting does not amount to any genuine conceptual difference.
67. Negarestani, "The Labor of the Inhuman," 452.
68. Negarestani, "The Labor of the Inhuman," 455.

constructivism: Norms are not found "out there" (in the universe) but are valorized only through our commitment to them. Accordingly, it is the task of "autonomous" reason to not only reimagine the world, but also to reconceive what it means to be a human being.[69] This existentialist ethos is one of "choosing for all of humanity," and indeed, defining what a human being is through one's freely chosen actions. It is Sartre's conception of *anguish*, that is, when a "man who commits himself [...] realizes that he is not only the individual that he chooses to be, but also a legislator choosing at the same time what humanity as a whole should be."[70]

The accelerationist merely takes this line to its logical conclusion. If we freely legislate for ourselves and for all humanity, beyond any preconceived limits, then we need not be limited by any set definition of what is "human." "Reason's main objective is to maintain and enhance itself. And it is the self-actualization of reason that coincides with the truth of the inhuman."[71] Hence "rationalist inhumanism" is the core philosophy of Left-accelerationists such as Reza Negarestani and Peter Wolfendale. It involves using reason to radically transform ourselves into something heretofore inconceivable and alien. For reason just is that supermundane faculty of "making possible certain abilities otherwise hidden or deemed impossible."[72]

Of course, one might rightly ask why such a project would be advisable in the first place? What is good about becoming "alien" or sublimely dissimilar from what we know of humanity today? One may respond to such objections by simply referencing the news or current events, and by pointing out the myriad ways in which we live an exploited and stultified existence under contemporary capitalism. Still, pointing out present degradations is clearly no defense for a faithful leap into the unknown. Whatever is "beyond" is as likely to be a dystopia as a utopia.

But this calls to mind an even more basic question for the accelerationists: How can we even begin to fathom our new, "inhuman" selves which we may hope to one day construct? Descartes makes a convincing point in his "Painter's Analogy," that devising an entirely new image—wholly disconnected from extant materials or ideas—would be strictly impossible.[73] All novel images are

---

69. Negarestani, "The Labor of the Inhuman," 438.
70. Jean-Paul Sartre, *Existentialism Is a Humanism*, trans. Carol Macomber (New Haven, CT: Yale University Press, 2007), 25.
71. Negarestani, "The Labor of the Inhuman," 437.
72. Negarestani, The Labor of the Inhuman," 437.
73. René Descartes, *Meditations on First Philosophy: With Selections from the Objections and Replies*, ed. and trans. John Cottingham, Revised Edition (Cambridge: Cambridge University Press, 1996), AT VII 19–20.

merely the recombination of more basic, already-existing elements. And yet, the accelerationist's voluntarism demands the creation of something pristinely new. For this, ordinary perceptions and mere understanding must be insufficient. What is needed is the imagination, and not only this, but an otherworldly conception of the imagination which can rupture our present realities.

> One cannot reproach a rational project for its phantasmatic residues unless one is secretly dreaming of a rationality that would be wholly devoid of imaginary influences. Prometheanism promises an overcoming of the opposition between reason and imagination: reason is fuelled by imagination, but it can also remake the limits of the imagination.[74]

As we have seen, what *appears* to be a judicious partnership is merely the indiscernible conflation of reason and imagination: they morph into one, supermundane faculty. For there is no priority of reason over the imagination, and hence no meaning to the term "reason" itself. Unless reason is understood as the sovereign, objective standard by which we choose, it has no propositional content whatsoever. It just *is* the will, or, our creative faculty. Reason is nothing apart from fiction-making. As Brassier most explicitly puts it at the beginning of the above excerpt: "Everything is more or less phantasmatic."[75]

Hyperstition

Hence, voluntarism naturally implies the third key concept within accelerationism: hyperstition. If the free will not only determines how we will act in the present, but also invents the future, then belief has priority over objective circumstances. *We* decide what the future will be. Not only this, but the future also has priority over the present. Here we come to the core feature of hyperstition, namely, that our chosen beliefs about the future (however fanciful) can retroactively form and shape our present realities.

To be sure, the humblest interpretation of "hyperstition" will be readily accepted by nearly everyone. If all that hyperstition means is that our beliefs, ideals, and goals for the future can affect how we organize society today, then this is nearly a self-evident statement. Even present-day institutions, such as marriage, the economy, and civil laws are essentially belief systems which gain their causal power through a community's common assent. The "full faith and credit" of the US federal government, which backs up treasury notes and

---

74. Brassier, "Prometheanism and Its Critics," 487.
75. Brassier, "Prometheanism and Its Critics," 487.

bonds, is a commonplace example of how beliefs can have material (and even financial) import.

If this is all that hyperstition means for the accelerationist, it would be a concept hardly worth talking about. Take the economy as an example: Undeniably, our faith in printed currencies and the inviolability of contracts have real-world implications, and the power of such beliefs is made evident (by way of contrast) in every failed state and economic depression where such common faith breaks down. Nonetheless, it is also clear that such beliefs do correspond to some material realities. "Money" is not only a collective delusion but is also a signpost of the real value stored in a given commodity. A commodity would have no value unless a certain amount of labor-time went into making the thing. Finally, labor involves particular concrete conditions which do not melt away before the supernatural force of the will. It requires, depending on the task, certain raw materials, tools, and training which cannot simply be "believed" away.[76] And yet, it is precisely this full-throated definition of hyperstition (that things can be believed in or out of existence) that the accelerationist must ultimately adopt. For them, the economy is truly a "hyperobject."[77] Belief not only expands upon material circumstances, but actually has the power to negate and supersede our circumstances altogether.

One might object here that our characterization of the accelerationist position is uncharitable. Material circumstances seem to matter with them. The accelerationist's preoccupation with innovation and technology strikes a materialistic tone. However, our foregoing discussion should make it clear that the accelerationist is no materialist. Their basic nominalism and voluntarism entail that there is a plurality of materials but no *matter*. As such, there are no natural laws common to all matter. It is truly up to the free will to define our circumstances and reorganize materials as we please. Far from technology

---

76. It is fashionable to mock the labor theory of value (i.e., that a commodity has value because it is worked on), as a relic of the nineteenth century, especially among economists who prize mathematical beauty over conceptual rigor and basic coherence. But the fact remains, all the same, that in the absence of labor, there would be no exchange values. We could make use of some of the fruits of nature without labor, but there could be no commodities. As Marx puts it in his correspondence to Ludwig Kugelmann, "Every child knows that a country which ceased to work, I will not say for a year, but for a few weeks, would die. Every child knows, too, that the mass of products corresponding to the different needs require different and quantitatively determined masses of the total labor of society"; Paul M. Sweezy, "Marx's Letter to Kugelmann," in *The Theory of Capitalist Development: Principles of Marxian Political Economy* (New York: Monthly Review Press, 1970), 41.
77. Reed, "Seven Prescriptions for Accelerationism," 529.

being the antithesis of fiction-making, both are equally the products of a free, creative will.

> Speculative possibility is effectuated through fiction, a fiction that maps vectors of the future upon the present [...] This is a fiction driven by anticipation (the unknown); a fiction that lacerates and opens the subject towards what awaits on the periphery of epistemic certainty. It is in this image that Accelerationism must embrace the fictional task of fabulating a generic will with a commitment equal to that which it makes to technological innovation.[78]

Here, those in the accelerationist camp will frequently try to outdo one another—criticizing their counterparts for being too empiricist and positivist, and for not paying sufficient respect to the role of belief. For example, Patricia Reed warns the authors of the #ACCELERATE Manifesto that they are in danger of being too pragmatic, and of neglecting the imagination. While encouraged by the "resurgence of ratiocentric discourse," Reed wants to play both sides. She counsels accelerationists to co-opt the theological tactics of the traditionalist Right for Leftist ends.

> Yet to embrace a central tenet of the [#ACCELERATE] Manifesto that suggests we build upon the "success of the enemy" entails not just the establishment of counter-think-tanks or the redirection of algorithmic-economic production towards other ends, but also a learning from the successes of the theological itself, intertwined as it is with any project directed toward the inexistent.[79]

But for their part, the authors of the #ACCELERATE Manifesto (Alex Williams and Nick Srnicek) appear perfectly at home with the notion of hyperstition. Alex Williams's essay "Escape Velocities" is explicit in affirming hyperstition as involving "narratives able to effectuate their own reality through the workings of feedback loops, generating new sociopolitical attractors."[80] Indeed, hyperstition, understood as "fiction that makes itself real through time-travelling feed-back loops," is part of the DNA of accelerationism, present in the originary works of the CCRU in the mid- and late 1990s.[81]

---

78. Reed, "Seven Prescriptions for Accelerationism," 529–30.
79. Reed, "Seven Prescriptions for Accelerationism," 528.
80. Williams, "Escape Velocities."
81. Simon O'Sullivan, "Accelerationism, Hyperstition and Myth-Science," *Cyclops Journal: Contemporary Theory, Theory of Religion, Experimental Theory* 2 (2017): 14.

For Williams and Srnicek, hyperstition always had a political valence. It is this connection between political program and fiction-making which has marked much of Left-accelerationist discourse in general. The very notion of progress itself is redefined. Accelerationists reject the idea that we can know, ahead of time, what a more humane or advanced form of civilization will look like. Instead, progress means choosing your own adventure. The future is inherently open, and so the utopias we can create for ourselves are really a matter of our own free choice. As such, "progress must be understood as *hyperstitional*: as a kind of fiction, but one that aims to transform itself into a truth."[82] And it isn't merely the case that the future is open descriptively; it is also indeterminate in a normative sense. That is to say, there is no objective standard by which we can judge the merits of our future utopia (or even to distinguish it from a dystopia).

The authors of the #ACCELERATE Manifesto will likely balk at such a strong characterization. They do profess, after all, an unequivocal fealty to the Left, and to the values of universalism and emancipation. Nonetheless, the logic of their position entails that their utopianism is entirely indeterminate. For if our future goals modify our present-day desires, and if those goals are a matter of free choice today, then we end up in a vicious circle. Free choice and emotion mutually condition one another without ever being grounded in a stable criterion such as human nature or human flourishing.

> Finally, in affirming the future, utopia functions as an affective modulator: it manipulates and modifies our desires and feelings, at both conscious and pre-conscious levels. In all its variations, utopia ultimately concerns the "education of desire." It provides a frame for us, telling us both how and what to desire, while unleashing these libidinal elements from the bounds of the reasonable.[83]

Therefore, while Left-accelerationists may affirm notions such as universalism and emancipation, it is perfectly indeterminate what those signifiers ultimately mean, or how they should concretely play out in society. In this, the Left-accelerationist reveals the superficiality of their connection to radical Enlightenment values. Historically, Enlightenment rationalism championed the supremacy of the intellect over affect, and thus sought to replace a romantic feudalism and aristocracy with a rational political order of universal rights. One needn't look further than the iconic conservative Edmund

---

82. Srnicek and Williams, *Inventing the Future*, 75.
83. Srnicek and Williams, *Inventing the Future*, 140.

Burke, and his *Reflections on the Revolution in France* (1790), for a clear dichotomy between Left rationalism and the Right romanticism that he preferred.[84] Similarly, on the Continent, the reactionary Joseph de Maistre opined that political constitutions could not be sustained by reason alone, and that a theological element was indispensable for civilization.[85]

It should not be surprising that Right has perennially championed a politics of affect over reason. If one wishes to support hereditary privilege and traditional hierarchy, then a nonrational bulwark becomes necessary for such arbitrary beliefs. But if, on the other hand, one wishes to unite peoples across cultures, languages, and ethnicities under a rubric of human emancipation, then affective stories and theologies are more likely a roadblock than a stable foundation. As Spinoza argues, the intellect is everywhere the same even while tastes and passions change from locality to locality.[86]

That the Right makes better use of affect and fiction is no mere relic of the eighteenth century; it is perfectly evident in contemporary politics as well. The 45th president of the United States, Donald Trump, was well-known for his casual relationship with the facts. (The recounting of untruths boasted by the former president on Twitter and elsewhere is already well documented and need not be reproduced here.) More interesting is the way that fiction-making has always been something of a conscious policy by Trump. As far back as the 1980s, Trump's *The Art of the Deal* promotes the idea of "truthful hyperbole."[87] Strikingly similar to the notion of hyperstition, truthful hyperbole involves

---

84. Edmund Burke, *Reflections on the Revolution in France* (London: Penguin, 2004).
85. Joseph de Maistre, *Considerations on France*, trans. Richard A. Lebrun (Montreal: McGill-Queen's University Press, 1974), 91–98.
86. Spinoza, "Ethics," E2P38; E4P35. There have, of course, been perennial attempts at constructing a Left politics steeped in passions and useful fictions. Georges Sorel, for example, is known for advocating the notion of the "general strike"—not as a concrete political proposal, but rather as a mere idea about the future to incite people's commitment and passions right now. As he put it in *Reflections on Violence* (1908), "There is no process by which the future can be predicted scientifically, nor even one which enables us to discuss whether one hypothesis about it is better than another." Nonetheless, "the framing of a future [...] may be very effective." However, Sorel's emphasis on political passions and "social myths" over scientific political theory is well known to have inspired, not only anarcho-syndicalist movements, but also fascist movements as well—in particular, the Italian fascism of Benito Mussolini. Georges Sorel, "Reflections on Violence," in *Socialist Thought: A Documentary History*, ed. Albert Fried and Ronald Sanders, Revised Edition (New York: Columbia University Press, 1992), 355–63.
87. Donald J. Trump and Tony Schwartz, *Trump: The Art of the Deal* (New York: Random House, 1987), 58.

exaggerating one's own greatness in order to convince others of this "fact," thereby (hopefully) making it so.

Certainly, the contemporary Right's use of conspiracy theories and "alternative facts" is not limited to the Trump legacy alone. The so-called Dark Web, and message boards like 8Chan, are repositories for white nationalist, Alt-Right personalities and anonymous posters. But rather than being horrified by these accretions of irrationality, much of Left-accelerationism appears to suffer from an acute jealousy. Faced with a fractured environment of siloed audiences and social media echo-chambers, despair has set in. And so, rather than appeal to the masses through common reason, the conclusion is reached that one should instead ape the Right's tactic of inflaming the passions of one's own, narrow base.

According to the Left-accelerationist tract, the #AltWoke Manifesto "narrative is more important than facts [...] Traditional pedagogy will not work in this instance." They therefore counsel the Left to question its own long-standing commitment to reason, and to trade in political theory for questions of media relations and rhetorical tactics. "Why is culture more important than policy? Why weaponize memetics? What is 'trickle down ideology'? Why support hyperreality and normalization?"[88] In a similar vein, Mohammad Salemy and Flavio Rossi advocate a "Mythopoesis of the Left" and, in particular, for imitating the myth-making strategies of the conservative psychologist Jordan Peterson:

> The Left needs to break out of its cyber bubble and start to develop more effective campaigns to win the hearts and minds of those who are not already in its milieu. The Left has to learn to stretch the borders of truth, produce semi-real heroes and stop apologizing for small mistakes. Instead of having a purely negative, resented and hysterical representation of the right-wing's activities, it must engage and use effective viral strategies to disrupt its adversaries from the inside.[89]

This jealousy of Right-wing tactics, and of the Right's "liberation" from fact-based discourse, palpably risks an acceptance of reactionary content. The seemingly innocent talk of imagining your own utopia quickly turns into the darker language of appropriating post-fact culture. On this point, the #AltWoke Manifesto is rather explicit that the Left should mimic Right-wing

---

88. "#AltWoke Manifesto," *&&& Journal*, February 5, 2017, http://tripleampersand.org/alt-woke-manifesto/.

89. Mohammad Salemy and Flavio Rossi, "Overcoming Left's Mythopoetic Deficit," *&&& Platform: The New Centre for Research and Practice* (blog), March 12, 2018, https://tripleampersand.org/overcoming-lefts-mythopoetic-deficit/.

conspiracy theories: "The Left does itself a disservice by not making its own. Speak their [i.e., the Right's] language to make it compelling: 'Peter Thiel is a member of the Bilderberg Group!'"[90]

For all their attempts to escape their dubious origins in the CCRU, and the explicit "neo-reactionary" ethos of Nick Land, Left-accelerationism appears caught in this irresistible orbit. As Simon O'Sullivan put it, "there is something compelling about the mythos Land deploys, even if one disagrees with the politics. More generally we have proof here that mythos, including the mythos of Land himself, is as powerful as any reasoned argument (or, indeed, rational programme)."[91]

There is something performative about all of this. Left-accelerationism learned from Nick Land the lessons of hyperstition and political fiction-making. They hoped to embrace the *form* of his politics while leaving behind its reactionary *content*. In other words, they want to be futurists without becoming fascists. But it is clear that hyperstition does not admit of any stable criteria by which to reject the myths of racial supremacy and Lovecraftian mysticism (i.e., Land's irrational program of a "return to the Old Ones").[92]

To the contrary, form is entirely bound up with content. A belief in the super-natural ability to choose, to act, and to create beyond all limits is inherently hostile to an ethos of universal solidarity. For given this premise, literally nothing binds us together in any meaningful way. We each create our own private utopias (despite the rhetoric of universal emancipation). For "free creations" are just not the sorts of things which can be made to cohere or align with one another. Neither, from these accelerationist premises, can it be presumed that all individuals (or even all peoples) are equally good creators. Land certainly doesn't think so—opining that white people need to preserve their love for liberty and innovation by "exiting" from multiracial democracies.[93]

Hence, accelerationist premises accommodate themselves far more naturally to an atomized individualism, and even a belief in racial hierarchies, than they do to a politics of universal solidarity and welfare. Imagination dovetails with particularisms of all sorts. That is why escaping the orbit of Nick Land is so hard for the Left-accelerationists to pull off. Left-accelerationism, despite the genuine political intentions of its adherents, is an oxymoron.

---

90. "#AltWoke Manifesto."
91. O'Sullivan, "Accelerationism, Hyperstition and Myth-Science," 29–30.
92. O'Sullivan, "Accelerationism, Hyperstition and Myth-Science," 13.
93. Land, "The Dark Enlightenment"; see also Harrison Fluss and Landon Frim, "Behemoth and Leviathan: The Fascist Bestiary of the Alt-Right," *Salvage*, no. 5 (October 2017), https://salvage.zone/in-print/behemoth-and-leviathan-the-fascist-bestiary-of-the-alt-right/.

Conclusion: The Inhumanity

As disturbing as these reactionary implications may be, they are only the penultimate consequence of accelerationist thought. Taken to its logical extreme, accelerationism not only destroys the "Cathedral" of multiracial democracy, but propels us to a totally alien and dizzying landscape. But this is no salvation, as though we could push past a nightmare by further losing control of our senses. Accelerationist premises finally lead to the most caustic departure from the human, or to use Brassier's early phrase, "the liquidation of man."[94]

For the accelerationist, the concept of the "human" is the ultimate subject of hyperstition. Hyperstition, as we have seen, is the idea that our beliefs about the future can return to change our present realities. In Reza Negarestani's formulation, "humanism […] is the initial condition of inhumanism as a force that travels back from the future to alter, if not completely discontinue […] its origin."[95] Humanity is thus a self-negating concept which carries the seeds of its own destruction.

The basic component of this inhumanism is the etherealization of "intelligence." Intelligence, or reason, must be made distinct from the carnal, lived experiences of *Homo sapiens* as we know them today. In a 2005 essay, Brassier complains that the conservative "defenders of reason" wish only to "maintain the honor of man against the philosophical incursions of 'scientism.'"[96] Accelerationists must boldly embrace the scientific image, even at the cost of destroying our comfortable, parochial self-images.

> But intelligence is neither reasonable nor sensible nor lived. It is time for philosophers to choose between carnal reasoning and "excarnate" intelligence. In selecting the latter we would prevent the spontaneous act of self-defense by which philosophers have tried to effect a transcendental immunisation of man against the contagion of an intelligence aiming for its release from the shackles of the human.[97]

This valorization of reason above traditional conceptions of the human *appears* closely aligned with Radical Enlightenment values. It reads as the supremacy of speculation over received wisdom and parochial "common sense." However, the accelerationist's notion of reason is, itself, entirely antinomian. Self-contradictory as this may seem, they view reason as infinitely

---

94. Brassier, "Liquidate Man Once and for All."
95. Negarestani, "The Labor of the Inhuman," 444.
96. Brassier, "Liquidate Man Once and for All."
97. Brassier, "Liquidate Man Once and for All."

malleable and revisable—an object of free commitment rather than a lawful mirroring of Nature. Accordingly, their reception of the Enlightenment is distorted in peculiarly anti-humanist directions, such that Brassier can speak of the "black dawn of the Enlightenment" as an event which will dissolve "mammalian stupidity" of *"la bête humaine."*[98] But for all of the disturbing imagery, the basic point is that "reason itself enjoins the destitution of selfhood" since it is no stable thing, but rather an irreducibly dynamic "act."[99]

Certainly, those in the Left-accelerationist camp surmise that this infinite revisability can be a springboard toward socialism and/or communism. They characterize capitalism as being conservative, and of maintaining this conservatism through its own traditional picture of the human. In their view, capitalism promises shiny new innovations and endless progress, but can only maintain itself through covert levers of control. Even this consumerist "progress" is boring, offering only new commodities to purchase in pedestrian strip malls. And while commodities and markets may change, the disciplinary regime of bourgeois norms (e.g., the authority of the boss, the doctor, and the father) remains all the same. Despite neoliberal claims of infinite dynamism, capitalism can maintain itself only if we remain obedient citizens and loyal brand-consumers.

For the Left-accelerationist, such conservative discipline is bound up with humanism. "Capital's human face is not something that it can eventually set aside, an optional component or sheath-cocoon with which it can ultimately dispense."[100] *Pace* Foucault, the Left-accelerationist argues that capitalism uses biopower in order to maintain consistency, consensus, and stability underneath a veneer of innovation.

To escape the rigors of capitalism, therefore, accelerationists insist on constantly upgrading what it means to be human. "Inhumanism" is not only an idea, but a revolutionary force unto itself. It "registers itself as a demand for construction: it demands that we define what it means to be human by treating the human as a constructible hypothesis, a space of navigation and intervention."[101] This project is committed to the idea that what it means to be human is "radically revisionary."[102]

But if the human is "subject to endless reengineering," then it is easy to see where this process ends.[103] Not only capitalism, but humans themselves, will

---

98. Brassier, "Liquidate Man Once and for All."
99. Brassier, "The View from Nowhere," 10.
100. Mark Fisher, "Terminator vs. Avatar," in Mackay and Avanessian, *#ACCELERATE#*, 345.
101. Negarestani, "The Labor of the Inhuman," 427.
102. Negarestani, The Labor of the Inhuman," 436.
103. Reed, "Seven Prescriptions for Accelerationism," 525.

need to be abolished. There can be no space for a hedonistic critique of capitalism, as failing to meet the conditions of human flourishing. For what is a human? What counts as flourishing? Or happiness? Rather than the "pleasure principle," it is the "death drive" which is most compatible with this process of endless revision. Everything, not only the cozy "idiocy of the village," but humans themselves, melt into thin air.

To posit reason as impersonal is to affirm this death instinct as pulling us "back towards the inorganic."[104] Rather than constituted subjects, with minds that mirror Nature, we are transformed into machinic parts for some superpersonal intelligence. We become inorganic cogs within a whole which no longer cares about our individual welfare or needs. Far from an indictment of an authoritarian capitalism, this is its speculative apotheosis. Humans are the mere "meat puppets" of the real agent of history, an ever-expanding capital or technology. Try as it might, accelerationism cannot escape the dark logic of Nick Land: "Man is something for it [capital] to overcome: a problem, drag."[105] Thus, inhumanism cannot sustain a revolutionary politics, but instead makes room for human subjugation. It requires no special prescience to see that the "liquidation of the human" is a prelude to the "liquidation of human beings."

---

104. Brassier, *Nihil Unbound*, 235.
105. Land, *Fanged Noumena*, 446.

Chapter 3

ECO-PESSIMISM

As with the previous discussion of accelerationism, the goal of this chapter will be to outline the contours of a particular worldview, in this case, what we call "eco-pessimism." It will then be possible to offer a critical analysis of this philosophical tendency. Here, we will maintain our focus on conceptual clarity instead of getting sidetracked by jargon and signifiers.

The first instance of this will be our use of the term "eco-pessimism" itself. The close-knit community of intellectuals discussed in this chapter hardly, if ever, apply this label to their own position. Nonetheless, the appellation describes well the actual orientation of this group, which includes, most notably, Donna Haraway, Isabelle Stengers, Bruno Latour, Déborah Danowski, Eduardo Viveiros de Castro, Dipesh Chakrabarty, and Clive Hamilton. It signifies their collective skepticism toward modern progress, neutral scientific authority, and the ability for human beings to endlessly reengineer their environments. Thus, the "pessimism" in eco-pessimism refers not to a broadly dour mood but more specifically to a suspicion about the possibility (and virtues) of technological progress.

One of the tropes of eco-pessimism is the constant derision of so-called geo-engineering, that is, the notion that one can significantly remake the landscape, including massive geological processes, in order to suit human needs and welfare. Eco-pessimists will often paint such grand projects as "eco-modernist," Promethean, and even accelerationist.[1] These writers favor the image of the Earth goddess, Gaia, over that of the titan Prometheus and the ethereal sky gods. In Bruno Latour's 2013 Gifford lectures, he sarcastically characterizes modernist ambitions:

> And there they are, seized by a new urge for total domination over a nature always perceived as recalcitrant and wild. In the great delirium

---

1. Bruno Latour, *Facing Gaia: Eight Lectures on the New Climatic Regime*, trans. Catherine Porter (Cambridge: Polity, 2017), 192, 240, 282. Predictably, their writing includes scorn for self-described Left-accelerationists, such as Alex Williams and Nick Srnicek.

that they call, modestly, geo-engineering, they mean to embrace the Earth as a whole. To recover from the nightmares of the past, they propose to increase still further the dosage of megalomania needed for survival in this world, which in their eyes has become a clinic for patients with frayed nerves. Modernization has led us into an impasse? Let's be even more resolutely modern![2]

For Latour, Gaia stands as an anti-modern, mythic figure. She is an image of the earth which defies our attempts to reorder nature as we please. Since "Gaia cannot be compared to a machine, it cannot be subjected to any sort of re-engineering."[3]

Gaian Metaphysics

Nonetheless, contemporary Gaianism is not reducible to this mythological signifier alone. Instead, it is composed of an entire constellation of positions when it comes to the nature of humanity, our relationship to the environment, and the definition of the good life. In some instances "Science" itself (understood as an authoritative, institutional complex) is taken up as a manifestation for man's hubris, and thus something to be opposed. As McKenzie Wark opines of the Gaian worldview, "Science becomes a kind of totem for Promethean man, able to brush aside all obstacles in the race to turn all of nature into a resource."[4]

Nonetheless, the Gaian position is not merely negative. While rejecting modern rootlessness and acceleration, the eco-pessimist promotes territoriality and "slow science."[5] As Latour succinctly puts it, "There is no cure for the condition of belonging to the world."[6] When the Gaians do seek inspiration from the scientific community, they often refer to James Lovelock, to whom the "Gaia Theory" is generally attributed. For it is Lovelock who most explicitly reversed the centuries-old tendency of "Science" to abstract from the local, rooted, and terrestrial "givenness" of the world.

The Gaians contrast Lovelock with Galileo Galilei, whom they see as the other great epoch-making scientist. While Galileo burst asunder our parochial,

---

2. Latour, *Facing Gaia*, 11, 12.
3. Latour, *Facing Gaia*, 96–97.
4. McKenzie Wark, *General Intellects: Twenty-One Thinkers for the Twenty-First Century* (London: Verso, 2017), 306–7.
5. Isabelle Stengers, *Another Science Is Possible: A Manifesto for Slow Science*, trans. Stephen Muecke (Cambridge: Polity, 2018).
6. Latour, *Facing Gaia*, 13.

geocentric perspective (affirming a heliocentric model), Lovelock affects an *Earth-centered* counterrevolution. Pointing to the static, dead atmospheres of other planets, he reinforced the uniqueness of the living Earth.

> We have to acknowledge that the symmetry is really too perfect: whereas the first scientist [Galileo] discovered how to shift away from the narrow view of the Grand Canal from his window toward the infinite universe, the second discovered how to shift from the infinite universe back to the narrow limits of the blue planet. What the first succeeded in doing with an inexpensive telescope, really a child's toy, the second accomplished by pointing an even lighter apparatus toward the sky—by performing a simple thought experiment.[7]

Human beings are thus no longer considered to be sui generis, autonomous entities. We do not exist apart from our particular "givens." Just as nature abhors a vacuum, so too does it abhor base abstractions. We exist only in the concrete, as part of a multispecies assemblage here on Earth. The desire to explore outer space, then, appears juvenile, if not altogether perverse. Nothing is "out there." Hence Latour's advice to our species at large: "Dream no longer, mortals! You won't escape into space. You have no dwelling place but this one, this narrow planet."[8]

But for all the criticism of "Science," technology, and space exploration, the Gaians seem reluctant to totally give up Enlightenment signifiers. Isabelle Stengers, for example, proudly calls herself a "daughter of the enlightenment."[9] She claims to problematize the Enlightenment's legacy, instead of discounting it altogether. Similarly, in Latour's *Down to Earth* (2018), he frequently asserts the need to think universally, and to devise improved forms of modernity and globalization.[10] Others, like Eduardo Viveiros de Castro and Déborah Danowski, are far less reserved

---

7. Latour, *Facing Gaia*, 75.
8. Latour, *Facing Gaia*, 81.
9. Isabelle Stengers, *In Catastrophic Times: Resisting the Coming Barbarism*, trans. Andrew Goffey (London: Open Humanities Press, 2015), 108.
10. Latour rejects what he calls the "modernization front," but does not want to be rhetorically associated with reactionary sentiments. Hence his distinction between "globalization minus" (i.e., the worst aspects of modernity) and "globalization plus," which would be a pluralistic world in which differences can flourish without modern forms of domination. See Bruno Latour, *Down to Earth: Politics in the New Climatic Regime*, trans. Catherine Porter (Cambridge: Polity, 2018), 14–15.

and insist that "the name of Gaia is an *anti-modernist* provocation" after all.[11]

What the eco-pessimists appear to agree on is that modernity is not as secular as it claims to be; instead modernity is just another example of religious faith. This rhetorical move should be familiar to all those acquainted with postmodern theory. Rather than rejecting modern values like objectivity and impartiality head on, the strategy is instead to deny the universal status of these values. They are a particular set of European, seventeenth- and eighteenth-century beliefs, no more universal than any other set of ethnic, religious, or cultural articles of faith.

If modernism is more dangerous than other belief systems, this is only because it cannot recognize its own particularity and situatedness. In perceiving itself as eternal and universal, modernity tends to colonize all nonconforming cultures and environments.

The Gaians specifically associate modern thinking with the ancient Christian heresy known as Gnosticism. While diverse in its origins (having Christian, Jewish, and Persian antecedents), Gnosticism came to be known for its rejection of corporeal matter as fallen. Accordingly, salvation was thought to be bound up with the pure spirit escaping bodily constraints. The political corollary for Gnostic thinking is often (perhaps ironically) the desire to achieve a *this-worldly* salvation. For while matter is considered to be irremediably fallen, salvation can be won, on this terrestrial plane, through advanced knowledge (*gnosis*). This heresy, therefore, sees the free and unbridled intellect as the true path to universal redemption.

That is why Latour casts modernity as just another faith, albeit a dangerous and heretical one which separates individuals from their lifeworlds. "If modernity were not so deeply religious, the call to adjust oneself to the Earth would be easily heard." Latour's criticism is here heavily influenced by the arch-reactionary Eric Voegelin.[12]

For the cold warrior Voegelin, his primary target was Marxism which he saw as a rationalist attempt to reengineer the world according to inflexible, abstract principles. Similarly, Latour chides the moderns for imposing their lofty "Ideals" upon corporeal reality. When such grand plans inevitably fail, the modernist exculpates the intellectual subject, and instead places all blame on the recalcitrant objects themselves.

---

11. Deborah Danowski and Eduardo Viveiros de Castro, *The Ends of the World*, trans. Rodrigo Nunes (Cambridge: Polity, 2017), 118.
12. Latour, *Facing Gaia*, 205–6.

By seeking to achieve Paradise on Earth, one succeeds only in realizing Hell on Earth—not always for oneself, but certainly for others. The failure of these projects—religious, scientific, technological, revolutionary, economic, governmental, the adjective hardly matters—leads those disappointed in Gnosticism to scorn matter even more, for its inability to rise to the level anticipated by the Ideal. Hence the strange position of objects, conceived at once as the sole reality and as the target of the deepest scorn.[13]

For Danowski and Viveiros, the apotheosis of this modernist, Gnostic heresy is to be found in accelerationism with its "Eurocentric eschatology of Progress […] nostalgic of a rationalist, imperialist, triumphalist past."[14] If there is to be a remedy for modernism, it is to be found in a genuinely *political* ecology. Such a political orientation will pay due respect to the bodily, the corporeal, and the material. It will, in other words, be "an object-oriented politics" based not in the autonomous, free subject (a mere chimera), but rather in specific territories, multispecies alliances, and soils.[15]

While raising the banner of "politics," this is nonetheless a politics which de-emphasizes human actors. Its anti-humanism is tied to the demotion of classic human qualities, such as self-consciousness and autonomy, and a general denial of the environment/subject distinction. In Clive Hamilton's words, "If humanism is the 'triumph of consciousness over its surroundings' […] we now witness the triumph of surroundings over consciousness."[16]

Eco-pessimists reject casting Gaia as a universal, comforting mother. Instead, she must be thought of as eminently political; she takes sides, and perhaps, even revenge. Gaia is opposed to the Gnostic moderns in the most resolute terms. "The great huntress of Gnostics […] Gaia is finitude, a very just and very worldly finitude."[17] It seems that Latour here associates Gaia with the "huntress" Artemis. Artemis is not merely a warrior, but also a protector of animals. Critically, she is linked with the quality of ecological *retribution*, doling out punishment during the Trojan War for the killing of her sacred deer.[18]

---

13. Latour, *Facing Gaia*, 208–9.
14. Danowski and Viveiros de Castro, *The Ends of the World*, 114.
15. Latour, *Down to Earth*, 52.
16. Clive Hamilton, "When Earth Juts Through World" (The Situation Facing the Moderns after the Intrusion of Gaïa: A Philosophical Simulation, Amphitheatre Caquot, Sciences Po, Paris, 2014).
17. Latour, *Facing Gaia*, 289.
18. George A. Christopoulos, *The Archaic Period* (Athens: Ekdotike Athenon, 1971), 108–10.

In pushing past the (often mutually conflicting) signifiers within Gaian literature, we at last come to the conclusion that their fundamental worldview is deeply antagonistic to both "materialism" and "rationalism," at least in the Enlightenment sense of those terms. The eco-pessimists are, moreover, united in denying any sense of universal Reason which transcends particular, physical, situated conditions. As Latour emphatically puts it, "If you want to keep using the words 'rational' and 'rationalists,' go ahead, but then also do the work of conceiving of the fully furnished spaces in which the presumed inhabitants can breathe, survive, equip themselves, and reproduce."[19] Nothing, not even reason, is beyond "the situation."

Superficially, Latour seems to be making a very conventional, materialist point. Reason is not a disembodied spirit, but is pegged to different configurations of materiality. However, Latour's position is not that of modern materialism as found in the writings of Spinoza, d'Holbach, or Marx. He rejects the materialist account of reason as a universal *logos* or set of physical laws (what Spinoza calls "*Natura naturans*").[20] Instead, for Latour, reason changes from place to place. Hence the import of the specific "territory" or "soil" as opposed to any uniform notion of "Nature."[21]

Likewise, this reveals how the Gaians approach the question of "materialism." They valorize objects themselves beyond their participation in uniform, self-same matter. In doing so, finite things gain a relative independence; they take on the qualities of intentionality and agency. This is the opposite of the Enlightenment view which sees finite things as always moved from the outside.[22] The Gaians reject this view as part of a degraded, mechanistic outlook. Instead, they counterpose an animism of spontaneous, self-movement and self-organization of objects.

---

19. Latour, *Facing Gaia*, 124.
20. Spinoza famously affirms that God's (i.e., Nature's) decrees are always uniform and universal. While the vocabulary sounds theological, this is really part of his rejection of miracles and divine intervention. To be more precise, the function of "natural laws" is understood through Spinoza's concept of the "infinite modes." These are the general features of the universe and are part of "Natura naturans" insofar as they govern all particular things and transient relations, namely, "Natura naturata." See Landon Frim and Harrison Fluss, "Substance Abuse: Spinoza Contra Deleuze," *Epoché: A Journal for the History of Philosophy* 23, no. 1 (2018): 191–217.
21. In fact, the Gaians typically reject the word "Nature" for precisely this reason. Latour, *Facing Gaia*, 226.
22. "A body which moves or is at rest must be determined to motion or rest by another body, which has also been determined to motion or rest by another, and that again by another, and so on, to infinity." Spinoza, "Ethics," E2 Physical Digression, Lemma 3.

All this is to say that corporeal entities, since always self-moving and spontaneous, cannot be subsumed under any more general laws of matter. This animistic (and thus anomalous) universe vexes the intellect. As Latour himself states, matter is neither an "inert object" nor a "higher arbiter."[23] "Gaia" stands opposed to "Nature" in precisely this sense. She is not the early modern concept of a total, intelligible substance, nor the theistic concept of a sovereign, omnipotent ruler. The question remains, however, as to what (or whom?) Gaia truly is.

Defining Gaia

In some sense, the cagey way in which eco-pessimists avoid defining Gaia is performative. Central to their worldview is the belief in a living, animate being who, nonetheless, eludes our best efforts at definition. To put it in a religious register, this approaches a negative, or "apophatic," theology. To the eco-pessimist, Gaia is a divine figure who can inspire awe but never a comprehensive understanding; she elicits feelings of "unpredictability, unfathomability, and a sense of panic in the face of a loss of control."[24]

This lack of total comprehension is tied to the object-oriented philosophy of the Gaian.[25] In negating any hard subject/object distinction, objects are foregrounded and gradually take on the traditional qualities of subjects. There is a horseshoe effect at work: Subjects are never truly banished from an object-oriented philosophy. Something must always fill the gap of agency, intentionality, and will. Rather than a universe populated by ordinary objects, everything becomes sublimely subjectified.

---

23. Latour, *Facing Gaia*, 280.
24. Danowski and Viveiros de Castro, *The Ends of the World*, 80.
25. The Object-Oriented Ontology (OOO) theorist Timothy Morton is occasionally suspicious of Gaia as a totalizing concept. Nonetheless, the way Latour and Stengers characterize Gaia seems rather compatible with what Morton has called a "hyperobject." Hyperobjects are "things that are massively distributed in time and space relative to humans." They, moreover, cannot be easily quantified by the sciences. Gaia, similarly, differs from ordinary, middle-sized objects typically treated by standard model physics. Most recently, Morton seems to have revised his views on Gaia, with Covid-19 cast as Gaia's emissary. See Timothy Morton, *Humankind: Solidarity with Non-Human People* (London: Verso, 2017), 3; Timothy Morton, *Hyperobjects: Philosophy and Ecology after the End of the World* (Minneapolis: University of Minnesota Press, 2013), 1; Timothy Morton, "Thank Virus for Symbiosis," April 4, 2020, https://www.thealternative.org.uk/dailyalternative/2020/4/4/corona-takes-roy-morton-lent.

Thus, the Earth is not merely another planet, but "a subject, and a planet like no other."[26] Gaia's having subjective qualities, however, is exactly why the eco-pessimist denies our ability to thoroughly know her. Subjects are animate, always moving, changing, and guided by an inner intentionality, rather than external, mechanistic laws. Therefore, "[s]he is always an antecedent and contradictory figure."[27]

As a subject, Gaia is not a totality. This point is essential, but easily missed by those who are accustomed to a second-wave feminist appreciation of Gaia, for example, by Marija Gimbutas and her followers. "Holism" used to be an inspiring adjective; it is now a bad word. While Gimbutas's generation of Gaians emphasized "Mother Earth" as a unified Creatrix, eco-pessimists today are far more circumspect. They insist that Lovelock's Gaia Theory never affirmed the notion of a unified planet or of Gaia as a superorganism.[28]

Gaia is now said to be an "outlaw" and an "anti-system." She is anomalous precisely because she is not engineered from mechanical parts, but instead composed of animate beings. Gaia is neither an inert creation nor a master creator, an engineer, or a clockmaker. As a result, she can neither be fully understood nor "re-engineered" according to human designs.[29]

All of this means that Gaia has for herself a very specific sort of will. Just as Gaia is irreducibly multiple, so is her sense of Providence.

> With Gaia, Lovelock is asking us to believe not in a single Providence, but in as many Providences as there are organisms on Earth. By generalizing Providence to each agent, he insures that the interests and profits of each actor will be *countered* by numerous other programs. The very idea of Providence is blurred, pixelated, and finally fades away. The simple result of such a distribution of final causes is not the emergence of a supreme Final Cause, but a fine *muddle*. This muddle is Gaia.[30]

The strategy here appears to be a renaturalization of the Gaia concept, only along nonrationalist lines. Instead of denying the ghostly teleology of

---

26. McKenzie Wark, "Bruno Latour: Occupy Earth," *Verso Blog* (blog), October 5, 2017, https://www.versobooks.com/blogs/3425-bruno-latour-occupy-earth.
27. Wark, "Bruno Latour." These philosophical moves are somewhat ironic. For the eco-pessimists reject not only seventeenth-century mechanism but also twentieth-century existentialism. The latter places a great deal of emphasis on the free and autonomous human subject. Existentialists, most notably the early Sartre, likewise move from the premise of a radically free human will to the indefinability of the human essence (at least a priori). See Sartre, *Existentialism Is a Humanism*.
28. Latour, *Facing Gaia*, 102.
29. Latour, *Facing Gaia*, 87, 96–97.
30. Latour, *Facing Gaia*, 100.

Mother Nature, one simply exaggerates and multiplies it to such an extent that all intentional plans cancel each other out. Of course, this is not a genuinely naturalistic ontology; blurring a multitude of volitions is not the same thing as affirming impersonal natural laws. But this is Latour's main point: The Gaian always aims for "an *addition* and not a *subtraction* of agents."[31]

This multiplicity of agents, where "the whole is inferior to the parts," has the effect of politicizing Gaia: She is not a passive, orderly Being, but a composition of agendas, intentions, and strivings. Irreducibly political, "Gaia is indeed a third party in all our conflicts."[32]

Stengers similarly describes Gaia as having agency, but no unified, normative agenda. Gaia is the one who "intrudes" and takes specific positions, often those hostile to human wellbeing. At the same time, she is also "indifferent" to specific human actors. The subsistence farmer living on a floodplain is equally (likely more) affected by climate change than the oil company executive living in a penthouse.

> Gaia is ticklish and that is why she must be named as a being. We are no longer dealing (only) with a wild and threatening nature, nor with a fragile nature to be protected, nor a nature to be mercilessly exploited. The case is new. Gaia, she who intrudes, asks nothing of us, not even a response to the question she imposes. Offended, Gaia is indifferent to the question "who is responsible?" and doesn't act as a righter of wrongs.[33]

In this, Stengers walks the same naturalistic tightrope as Latour. Gaia is a subject who, through a transcendent indifference, acts like an implacable object at the same time.[34] She is but "an assemblage of material processes that demand neither to be protected nor to be loved, and which cannot be moved by the public manifestation of our remorse."[35] Yet again, this is no genuine naturalism. Instead, what we find in each of these writers is the desire to sublimate political agonism into an incomprehensible pluriverse. They avoid sounding like traditional theists only by multiplying Providences, damnations, divine edicts, and political decrees. But a polytheism is a theism nonetheless.

The practical result of politicizing Gaia is that the entire project of modernity is called into question. If Gaia is, herself, a political actor, then she can no longer be conceived as an inert pool of resources to be exploited or managed. From the eco-pessimists' point of view, the climate crisis we find

---

31. Latour, *Facing Gaia*, 163–64.
32. Latour, *Facing Gaia*, 238.
33. Stengers, *In Catastrophic Times*, 46.
34. Wark, *General Intellects*, 300.
35. Stengers, *In Catastrophic Times*, 48.

ourselves in is the result of our own hubris—our literally toxic practice of deanimating nature and overestimating our own agency.

> They [moderns] have allowed this state of affairs because they are unwilling to control themselves. They will not turn down the music at the party of progress. So their technology has taken on an autonomous power. Now a titanic struggle brews. It is not, as the Moderns always imagined, Man *versus* Nature but Technology *versus* Gaia.[36]

It is clear that technology is the concrete manifestation of man's hubris. Prometheus's fire (that archetypal symbol of techne) is not only a rebellion against established order (Zeus), but is also a rebellion against nature, that is, Gaia herself. Here, then, we have a literally titanic clash of worldviews, a genuine *titanomachy*.

The Promethean outlook is defined by three, interlocking ideas. These are "Nominalism," "Voluntarism," and "Hyperstition." In other words, the world is so insubstantial, so atomized and lacking in stable essences, that it can be voluntarily remade according to a Promethean will. In this scheme, the only limit to our creativity is our own "hyperstitial" imaginations.

In contrast to this, the Gaian worldview is defined by the three ideas of "Noumenalism," "Animism," and "Humility." The world does have a deep, significant meaning which is beyond human grasping. Gaia is inscrutable because she is alive, animate, and constantly on the move. It is not up to us humans to change or reengineer her, but rather to live humbly according to her edicts.

Noumenalism

Noumenalism can mean a lot of different things depending on the philosophical context. But in all cases, it is a statement about the insuperable limits to knowledge. Eco-pessimists tend not to use this word in describing their own theories. Yet, it is perfectly evident that a fundamental (perhaps *the* fundamental) attribute of Gaia is her transcendence of human understanding. As Clive Hamilton so aptly put it, "If Gaia exists, why would she debase herself to the level of human understanding? [...] At best, peering at Gaia illuminated by the light of reason can reveal no more than Gaia's silhouette. Her

---

36. Clive Hamilton, "Gaia Does Not Negotiate" (The Situation Facing the Moderns after the Intrusion of Gaïa: A Philosophical Simulation, Amphitheatre Caquot, Sciences Po, Paris, 2014), https://clivehamilton.com/gaia-does-not-negotiate/.

enigma is her essence. So he who says he has solved the enigma of Gaia is deluded or lying."[37]

Some of the more circumspect eco-pessimists, notably Dipesh Chakrabarty, will make moves in the direction of universalism. Gaia is a figure which can, in some limited sense, be experienced by diverse peoples. But even here, there are strict limits placed on our capacity to know. A phenomenon such as climate change is experienced the world over, but can scarcely be "understood" in any rigorous, final sense. It can never be subsumed within a Hegelian or dialectical logic, but is always "catastrophic"—both physically and in an epistemological sense:

> We can never *understand* this universal. It is not a Hegelian universal arising dialectically out of the movement of history [...] Yet climate change poses for us a question of a human collectivity, an us, pointing to a figure of the universal that escapes our capacity to experience the world. It is more like a universal that arises from a shared sense of a catastrophe.[38]

Crucially, this limit to human understanding is not only descriptive for the Gaian, but also moralistic. To seek full knowledge is not only impossible, but a sin of pride as well. Continuing his association of scientific rationalism with the Gnostic heresy, Latour points to the etymology of "Gnosticism" as "assured knowledge," and counterposes this with true faith. "Faith is what grasps you; knowledge is what you grasp."[39] In a similar vein, Donna Haraway links the presumption of knowing God's proper name with the sin of idolatry.[40] And for her part, Isabelle Stengers considers the imposition of a singular name, or understanding, of Gaia to be a "barbaric" act. Linking apophatic noumenalism with an emphasis on plurality, Stengers insists that any characterization of Gaia must be nonglobal, and composed through many hesitant, stammering voices:

> Naming Gaia is naming a question, but emphatically not defining the terms of the answer [...] Learning to compose will need many names, not a global one, the voices of many peoples, knowledges, and earthly

---

37. Hamilton, "Gaia Does Not Negotiate."
38. Dipesh Chakrabarty, "The Climate of History: Four Theses," *Critical Inquiry* 35, no. 2 (Winter 2009): 221–22.
39. Latour, *Facing Gaia*, 203.
40. Fabrizio Terranova, *Donna Haraway: Story Telling for Earthly Survival* (Brooklyn, NY: Icarus Films, 2018).

practices [...] There will be no response other than the barbaric if we do not learn to couple together multiple, divergent struggles and engagements in this process of creation, as hesitant and stammering as it may be.[41]

It should not be thought that Gaian negative theology is merely an aesthetic preference or a philosophical "mood." Rather, there is a (more or less) consistent metaphysics at work here. Their unique brand of noumenalism is distinct from the standard, Kantian variety. In Kant's critical philosophy, the noumenal is defined in opposition to the phenomenal world. The latter is always patterned by the transcendental conditions of understanding, that is, the fundamental requirements for forming any thoughts at all.

Gaian noumenalism takes a more historicist turn, influenced by Latour's interests in sociology and anthropology. Instead of a veil of perception, separating the knower from the pristine *"Ding an sich,"* the phenomenal world itself becomes occluded. Here, the fundamental opposition is not between the noumenal and the phenomenal, but between "nature" and "culture." For Latour, one can never hope to pacify the relations between these two domains; nothing is entirely "natural," and one must always live with the unstable and undecidable interplay between the two categories. This has the overall effect of precluding final, comprehensive knowledge, but leaves open the possibility for local investigations. One can never jump the boundary between provisional understanding and true knowledge, but, as Latour puts it, "it is not entirely impossible to probe *on the near side*."[42]

This relates to Latour's sociological diagnosis of "blackboxing."[43] Simply stated, blackboxing is the social process whereby complex things are made to appear simple, monolithic, or unproblematic. Societies typically blackbox common machines and technologies. That is to say, we don't think too hard about how complicated equipment works, especially *when* it works. But this process could be applied to any number of objects, beings, or social institutions; all that is required is the desire not to think too critically about things. The idea of the "black box," then, presents an immanent sort of noumenalism. The occluded "thing-in-itself" is never beyond this world, but instead populates every facet of our own reality.

Individual entities are opaque to human understanding. This is because they are not engineered from discrete parts which, themselves, can be understood

---

41. Stengers, *In Catastrophic Times*, 50.
42. Latour, *Facing Gaia*, 21.
43. Bruno Latour, *Pandora's Hope: Essays on the Reality of Science Studies* (Cambridge, MA: Harvard University Press, 1999), 131.

in isolation. The more one unpacks a given entity, the more one realizes that an irreducible complexity runs all the way down. Hence Latour appreciates James Lovelock's insight that "there are neither [discrete] parts nor a whole."[44] But this post-structuralism also means that there are no neat borders between the "inside" and the "outside" of any given entity. In Haraway's words, "critters do not precede their relatings."[45] Objects are defined, instead, by "waves of actions" which are inextricably connected to other, proximate entities.

> The inside and outside of all borders are subverted. Not because everything is connected in a 'great chain of being;' not because there is some global plan that orders the concatenation of agents; but because the interaction between a neighbor who is actively manipulating his neighbors and all the others who are manipulating the first one defines what could be called *waves of action*, which respect no borders and, even more importantly, never respect any fixed scale.[46]

Absent of insides and outsides, borders, and fixed scales, the world necessarily eludes our total comprehension. One can never open up all the black boxes (to try would, in any case, be a sin of hubris). But at the same time, it would be equally wrong to take an uncritical, lazy attitude toward our environments and pretend as though the infinite, problematical complexity isn't really there. The answer (though Gaians hate the word "answer") is therefore to engage only in provisional, limited sorts of investigations. We must resist hasty moves to a "universal," "natural," or "global" perspective, and instead pay proper attention to the always situated, always mediated ways in which we gain partial answers.

> Understanding the entanglements of the contradictory and conflictual connections is not a job that can be accomplished by leaping up to a higher "global" level to see them act like a single whole; one can only make their potential paths cross with as many instruments as possible in order to have a chance to detect the ways in which these agencies are connected among themselves. Once again, the global, the natural, and the universal operate like so many dangerous poisons that obscure the difficulty of putting in place the networks of equipment by means of which the consequences of action would become visible to all the agencies.[47]

---

44. Latour, *Facing Gaia*, 95.
45. Haraway, *Staying with the Trouble*, 60.
46. Latour, *Facing Gaia*, 101.
47. Latour, *Facing Gaia*, 140–41.

Latour has special venom for the geometry of the "globe" or "sphere" itself, considering it to be the apotheosis of universal ignorance of the finite:

> Even so, it must be possible, today, to pull ourselves away from the fascination that the image of the Sphere has held for us since Plato: the spherical form rounds off knowledge in a continuous, complete, transparent, omnipresent volume that masks the extraordinarily difficult task of assembling the data points coming from all instruments and all disciplines. A sphere has no history, no beginning, no end, no holes, no discontinuities of any sort. It is not merely an idea but the very ideal of ideas.[48]

Gaian noumenalism is therefore a moving target. There is no end to investigation, no satisfying meta-narratives, no complete perspectives, but only infinite, irreducible complexity and interconnectedness. Unlike its Kantian counterpart, this sort of noumenalism can be pierced by human investigation, only never in a comprehensive or final way. The correct disposition to this worldview is one of "irony," "humor," and "serious play," at least according to Donna Haraway.[49] The world-historical resolutions of Hegel are displaced by the ironic grin of the wise rustic.

After all, in such a world, there can be no "end of history," but only a succession of totally novel experiences. In Latour's words, "We are idiots; everything that happens to us happens only once, only to us, only here."[50] Accordingly, we should be open to being surprised; monotony is but a modern illusion.

> History, for its part, surprises us and obliges us to start all over again every time. The impression of repeating the same thing comes from the form of the Globe with which everyone tries to depict what is happening to it that is new. In contrast, the discovery, shattering every time, of a dramatic new connection between previously unknown agents, and on increasingly more distant scales, and at an increasingly frenetic pace – yes, this is truly new.[51]

This advice about proper comportment again indicates the moralistic attitude of the eco-pessimist. (Indeed, the blurring of the fact/value distinction is

---

48. Latour, *Facing Gaia*, 136.
49. Donna J. Haraway, *A Cyborg Manifesto: Science, Technology, and Socialist-Feminism in the Late Twentieth Century* (Minneapolis: University of Minnesota Press, 2016), 5.
50. Latour, *Facing Gaia*, 81.
51. Latour, *Facing Gaia*, 138.

a constant theme in much of their writing.) Post-structuralism not only asserts the fact of the world's novel complexity, but also that instances of reduction, sameness, and uniformity are *normatively* bad. This is exemplified in the way Haraway speaks of environmental degradation: "Multi-species alliances" are good; diverse coral ecosystems are good; but one should have contempt for the monotony of overstressed environments.

Monotonous, "slimy," unvariegated habitats can only be the result of absorbing "insult after insult."[52] Danowski and Viveiros are even more straightforward in denouncing uniformity using starkly normative terms. They invoke Cormac McCarthy's novel, *The Road*, in speaking of the struggle "against the decomposition of the world into grey toxic mud."[53]

On the one hand, it may seem perfectly intuitive to characterize an achromatic, static environment, lacking in biodiversity, as a bad thing. We know that human beings are but a small part of a complex, interconnected web of biological and nonbiological entities which we depend upon for our very survival. This is beyond debate.

On the other hand, one may ask what the locus of moral concern is, when it comes to Gaian writing. Is environmental monotony merely bad in an indirect way (i.e., that this would immiserate the lives of human and nonhuman animals)? This is certainly true, but hardly a unique philosophical insight.

Or, alternatively, does the Gaian place *inherent* moral worth upon the quality of diversity itself? This, to be sure, would be a nontrivial philosophical claim—though, one which may be difficult to substantiate. Consider, for example, a distant exo-planet which contains no sentient life whatsoever. Would it really matter (in a moral sense) whether the species of grass on that world were of one, two, or twenty-seven different colors? What good is diversity apart from sentient appreciation?

It seems clear that the eco-pessimists do opt for the *intrinsic* valorization of plurality, novelty, and difference. Even in her earlier writings on the concept of cyborgs (i.e., the various interplays of organic life and machines), Haraway insists that such beings disdain to reflect the cosmos as an integrated, consistent whole. Cyborgs are said to be "not reverent." "They are wary of hol-ism, but needy for connection."[54]

Her position, at least in this regard, seems to have changed very little. Haraway's recent works tend to avoid consistent "universals" such as Nature, and instead emphasize the particularity of "terra's pluriverse." "Terrapolis is

---

52. Haraway, *Staying with the Trouble*, 56.
53. Danowski and Viveiros de Castro, *The Ends of the World*, 52.
54. Haraway, *A Cyborg Manifesto*, 9.

open, worldly, indeterminate, and polytemporal."⁵⁵ Again, this is a thoroughly normative and political statement. Her vision of multispecies society is one "shed of masculinist universals and their politics of inclusion."⁵⁶

Latour similarly counsels that we "agree to remain open to the dizzying otherness of existents," citing the "pluriverse" concept held by the American pragmatist William James.⁵⁷ Anticipating the COP21 conference held in Paris, Latour promoted his own mock version of what a truly pluralistic climate summit should look like. This meeting, which he characterized as a "cosmopolitical" process, included delegates, not only from the various nations of the world, but also representing certain nonhuman collectives (such as "Ocean," "Forest," "Atmosphere," and "Land"). The crucial thing, however, was that this diplomatic performance could only occur once each facet of the world was understood, *and respected*, as a unique, political player with its own particular interests and agendas.

> The members of the [simulated] COP are not *parts* of a higher Whole that would allow them to be unified by attributing to each a role, a function and limits; rather, they are "parties" in the diplomatic sense, in a negotiation that can begin precisely only because there is *no longer a higher arbiter*—neither power, nor law, nor nature.⁵⁸

The point, then, is that complexity is a normative good in itself, and so our social and political actions are likewise judged morally according to this standard. We act well when we respect differences and act poorly when we try to subsume all difference under hasty, universal constructs.

How does this post-structural valorization of difference bear upon the question of materialism? The Gaians all purport to be materialists at some point in their writings, although they claim that this is a "terrestrial" materialism as opposed to an abstract and rationalistic one. As Haraway describes her position alongside those of Stengers and Latour, they all affirm a materialism "committed to an ecology of practices, to the mundane articulating of assemblages through situated work and play in the muddle of messy living and dying."⁵⁹

Their ostensible materialism is also in contrast to "epic" or Promethean materialism, in which "nature is there as a resource for human conquest."⁶⁰

---

55. Haraway, *Staying with the Trouble*, 11.
56. Haraway, *Staying with the Trouble*, 12.
57. Latour, *Facing Gaia*, 36.
58. Latour, *Facing Gaia*, 259.
59. Haraway, *Staying with the Trouble*, 42.
60. Wark, *General Intellects*, 302.

Latour even makes a sharp distinction between "matter" and "materiality." For him, "matter" is a bloodless, universal, and abstract category that smacks of idealism, while "materiality"—that which is situated, "worldly," dense, and full of practical relations—is truly materialist.[61]

But here, we have to be careful. For the signifiers regarding matter and materialism become unreliable within Gaian discourse. In some instances, Latour will claim that "to be a materialist, one has to have matter."[62] In other places, he casts matter as a religious pseudo-concept, that is, part of a Gnostic subversion of concrete reality in favor of yet another abstract, universal concept.[63]

Whatever their chosen vocabulary, the Gaian worldview is relatively consistent when it comes to material reality itself. What they all seem to reject is the Enlightenment concept of *res extensa*, or in other words, uniform, extended space. Everything else they claim about matter or materiality follows from this primary rejection. This involves the denial that primary qualities (such as number, length, motion, shape, size, etc.) have any priority over what are traditionally considered to be the secondary qualities (color, scent, sound, etc.). The particularities of "texture, but also procreation, aging, and death" are just as fundamental and primordial to so-called matter as extension itself.[64]

This is (ironically for anti-humanists and object-oriented philosophers) a species of existentialism. In trying not to impose human categories of understanding onto matter, everything that happens in reality starts to *matter*, and indeed, to matter equally. The human life-world is as fundamental to the nature of existence as are the simplest laws of physics.

All irony aside, the upshot of this method is to deny the "billiard balls" model of the universe. As against "Kepler, Cyrano, Descartes, Fontenelle, and Newton," the universe is not populated by discrete objects that enter into external relations with one another, governed by universal and unchanging physical laws.[65] Following the "blackbox" noumenalism discussed above, this denial of predictable, external relations should apply to any collection of objects whatsoever. But at the very least, Latour denies the existence of discrete parts when it comes to the living world, that is, "Gaia."

> This is the sense in which there cannot be, strictly speaking, any parts. No agent on Earth is simply superimposed on another like a brick

---

61. Latour, *Down to Earth*, 60; Latour, *Facing Gaia*, 207.
62. Latour, *Down to Earth*, 60.
63. Latour, *Facing Gaia*, 207.
64. Latour, *Facing Gaia*, 77.
65. Latour, *Facing Gaia*, 77.

juxtaposed to another brick. On a dead planet, the components would be placed *partes extra partes*; not on Earth. Each agency modifies its neighbors, however slightly, so as to make its own survival slightly less improbable.[66]

If there are no discrete parts in the living world, then what is primary is change itself. This presumes a reversal of the typical relations between time and space. On the classic materialist model, temporal events depend upon a change in spatial relations. Matter is logically prior to the particular events in which it takes part. Even in Einsteinian physics, "space-time" links the two concepts within a calculable, predictive model.

The Gaian worldview, by contrast, "subverts all the levels." Since relationality and change are, themselves, fundamental, "*space is the offspring of time*" (i.e., becoming is prior to being).[67] Whatever materialist signifiers they may employ, the eco-pessimist is, at bottom, an idealist. This is for at least two reasons: First, everything on the Gaian model appears to be "ensouled" à la Leibniz's *Monadology*. Invoking Leibniz, Danowski and Viveiros proclaim that "there are souls everywhere."[68] Spontaneous self-movement comes before material identity or physical laws.

Second, since the Gaians deny any objective, scientific viewpoint, their reference is necessarily phenomenological and experiential. In classical Marxist terms, this means prioritizing consciousness over being, or the ideal over the real.[69] What is rejected, in any case, is the modern notion that science offers explanations according to "a chain of causal events, and these, in turn, to materially dense interactions."[70]

Therefore, Gaia not only "subverts the levels" of space and time, but of cause and effect as well. Indeed, the very notion that time "passes *from the past toward the present*" is considered, by Latour, to be a peculiar innovation of rationalism, rather than an unproblematic, analytic statement.[71] In this way, rationalism and classic materialism are said to illicitly deprive the world of its "drama" and "adventure." For nothing, on such a deterministic account, could truly be new or surprising.

---

66. Latour, *Facing Gaia*, 98.
67. Latour, *Facing Gaia*, 106.
68. Danowski and Viveiros de Castro, *The Ends of the World*, 71.
69. Friedrich Engels, *Ludwig Feuerbach and the End of Classical German Philosophy* (New York: International, 2018), 20–21.
70. Eduardo Viveiros de Castro, *Cannibal Metaphysics: For a Post-Structuralist Anthropology*, trans. Peter Skafish (Minneapolis: Univocal, 2014), 61–62.
71. Latour, *Facing Gaia*, 54.

Once more, there is this persistent denial of the fact/value distinction. Not only is a causally determined world boring, but it is also a factual (ontological) mistake. For Latour, such a world—which runs from the past, up through the present, and into the future—places all agency in causes and deprives effects of all power. This, he claims, is "obviously impossible." And why is this so impossible? Because, for Latour, this is simply not how one experiences the world.

> The consequences are always surprising and, in practice, in the history of discovery, as in the narrative of discovery, and even in the teaching of the most solidly established facts, the cause arrives a long time *after* the consequences [...] a strictly *causalist* narrative in which a single character, the sole actor, would be in the cause—and furthermore in the primary cause—is obviously impossible. By definition it would be impossible for anyone to produce such a narrative.[72]

This is a repetition of the ironic humanism seen above. The promotion of spontaneous, surprising matter is founded upon our own phenomenological experiences with scientific discovery. That is because we experience effects before their causes, therefore we conclude that effects are *really prior* to causes (in a metaphysical sense). Of course, this is a strikingly anthropocentric leap. It merely reifies the order in which we happen to experience things, rather than thinking critically about what these concepts actually mean or even if our experiences are isomorphic with reality itself.

Nonetheless, the eco-pessimist writes as though placing causes before effects is an artificial operation, peculiar to modern "scientistic" thinking. Once more, they tar and feather rationalism with the charge of being just another religious faith. As Latour puts it, "this form of causalist narrative closely resembles the *creationist* stories through which one attributes to a first cause, to a creation deemed *ex nihilo*, the whole series of what follows [...] It hardly matters, then, whether the antecedent is called an omnipotent Creator or omnipotent Causality."[73]

Despite his dismissal of determinism as "religious," however, Latour also derides the scientific worldview for its capacity to disenchant the world. In a reality governed by efficient causation, there are no true "events," but only mundane episodes which are reducible to their antecedent causes. "The great paradox of the 'scientific worldview' is that it has succeeded in *withdrawing* the *historicity* of the world for science as well as for politics and religion."[74]

---

72. Latour, *Facing Gaia*, 68–69.
73. Latour, *Facing Gaia*, 71.
74. Latour, *Facing Gaia*, 72.

Therefore, Latour (as with many of the other Gaians) adopts an odd rhetorical strategy: They disparage their rationalist opponents as *merely* believers in a particular faith, while also wanting to defend their faith in an enchanted world.

This enchantment of the Gaian world follows directly from their inversion of causes and effects. If effects are truly prior to their causes (whatever that could mean), then this triggers a whole series of subsequent reversals and collapses. For one, environments are no longer prior to the beings which populate them. Recall that this was the fundamental insight of James Lovelock, and the major reason why he is celebrated by the eco-pessimist tendency to this day. Living creatures make the world habitable for living creatures.

In Danowski and Viveiros's language: "the *ambiented* becomes the *ambient* (or "ambienting"), and the converse is equally the case. It is effectively the collapse of an ever more ambiguous environment, of which we can no longer say *where* it is in relation to us, and us to it."[75]

Danowski and Viveiros then connect this reversal to the metaphysics of certain Amerindian peoples. They promote the idea that nature *itself* is a plurality. The plurality of nature, in fact, *precedes* any culturally determined "way of seeing." In this, the reversal of "cause and effect" transforms into the reversal of "nature and representation." It is not that the world is merely *seen* in different ways by different cultures; to the contrary, nature is metaphysically plural according to the various beings which compose it.

They claim that, within Amerindian mythos, every being (indeed, every facet of nature) is a literal person, or more specifically, a "human." Whether they are biologically a *Homo sapien*, a tapir, or a jaguar makes no difference. In this "multinaturalism," each diverse perspective, itself, creates the natural world. As Wark explains their position:

> Amerindian myth has another interesting aspect. Here all beings—the pig, the jaguar, and the human—see the world the same way, but they see different worlds. There is not one nature and multicultural ways of seeing it. To the contrary, there is one way of seeing a *multinaturalism*. There is no thing-in-itself. It may be not so much a variety of natures, so much as nature as variation that the different "species" (as science would call them) perceive.[76]

Does this count as a genuine noumenalism? It does not, just in case one adopts the narrow, Kantian definition of noumenalism as containing a

---

75. Danowski and Viveiros de Castro, *The Ends of the World*, 14.
76. Wark, *Sensoria*, 139.

singular "thing-in-itself" beyond our multifarious perspectives. Viveiros, in his *Cannibal Metaphysics*, explicitly rejects that sort of noumenal picture.[77] But the eco-pessimists do affirm an even more radical version of noumenalism. This is consistent with Latour's notion of blackboxing, as well as the reversals of "cause and effect," "being and environment," "identity and difference," and "reality and perspective."

Amerindian perspectivism is less Kantian than it is Leibnizian.[78] Danowski and Viveiros explicitly compare individual perspectives with Leibniz's concept of "monads," that is, individual souls which compose the world.[79] We cannot know the thing-in-itself because it is always changing, coevolving, and plural; it always has a shifting point of view.

The noumenal, for the Gaian, is composed of *things* and is not a singular *thing*. This enchanted vision of the world tries to be idealist and materialist, naturalist and perspectival, all at once. It is as if the "chicken and egg" riddle could be resolved by simply inventing the new word, "chegg." But this is in keeping with their Leibnizian metaphysics where the entire world is inherently plural, perspectival, and intentional. This is true at the elementary level of the soul/body, and not applied after the fact in some intellectual operation.

> A perspective is not a representation because representations are properties of mind, whereas a *point of view is in the body*. The capacity to occupy a point of view is doubtlessly a power of the soul [...] The latter [i.e., the soul], being formally identical across all species, perceive the same thing everywhere. The difference, then, must lie in the specificity of the body.[80]

In this neo-Leibnizian picture, there are no final answers, but only an apophatic delay. One can never move beyond the plurality of perspectives because plurality, difference, and change are not matters of perspective at all—they are baked into the cake of existence itself.

Their negative theology then has practical effects for how one negotiates the world. Barring a stable, objective reality, and keeping with the inversion of "cause and effect," the best one can do is to search for concrete practices and behaviors that appear to work here and now.

In this, Isabelle Stengers makes great use of the concept of the *pharmakon*, as found in the Platonic dialogue, *Phaedrus*. The term, which is the root for the modern English words "pharmacy" and "pharmacology," originally had an

---

77. Viveiros de Castro, *Cannibal Metaphysics*, 73.
78. Danowski and Viveiros de Castro, *The Ends of the World*, 70.
79. Danowski and Viveiros de Castro, *The Ends of the World*, 72.
80. Viveiros de Castro, *Cannibal Metaphysics*, 72.

ambiguous meaning—referring both to a "cure" and a "poison." The philosophical point is that efficacy and pragmatic use are emphasized over knowledge of a substance's true nature or essence.

Put otherwise, it is far more important to determine what a thing can do, in this particular circumstance, than "what it is" in some transhistorical, universalist sense. In Stengers's words, "What characterizes the pharmakon is at the same time both its efficacy and its absence of identity. Depending on dose and use, it can be both a poison and a remedy."[81]

The "pharmakon" serves as a model for a particular view of political activism and policy-making. Since there are no final answers in this noumenal reality of ours, all policy must be provisional and negotiated from the "bottom up," instead of imposed by elite experts and big Science. All peoples (but especially those who have been historically marginalized) must be consulted. And whatever solutions are arrived at (e.g., concerning food policy) must be understood as merely provisional, and not necessarily applicable to other regions, cultural contexts, or times.

> There is no generality that would define a different agriculture, one that is able to compose itself with Gaia, but also to stop poisoning the concrete Earth and its inhabitants, and this whilst feeding ever growing numbers of humans. Not that this is impossible but the possibilities have to be formulated on a case by case, region by region basis, and above all in a mode that confers a crucial place to the knowledges of interested people [...] Multiple connections are to be created and maintained, never to be considered acquired once and for all.[82]

Being cognizant about geographical and historical contexts as well as paying due respect to marginalized peoples are certainly important when it comes to crafting effective policy. However, what Stengers is actually suggesting here goes far beyond an empirical appreciation of local differences. Rather, hers is a far-reaching, rigorous denial of all universals, including that of humanity itself.

> For my part, I have never encountered "people," only ever persons and groups, and always in circumstances that are not simply a context but

---

81. Stengers, *In Catastrophic Times*, 100. Stengers's discussion is strongly reminiscent of Jacques Derrida's interpretation of the *Phaedrus* dialogue. See Jacques Derrida, "Plato's Pharmacy," in *Dissemination*, trans. Barbara Johnson (Chicago, IL: University of Chicago Press, n.d.).
82. Stengers, *In Catastrophic Times*, 128.

which are operative. Thus, what interests me with the example of juries, whether in a court case or a citizen jury, is not that they would manifest the equality of humans when it is a matter of thinking. It is *the efficacy of an apparatus that brings about a "making equal."*[83]

Stengers's point here is wholly consistent with her broader pragmatism and focus on "efficacy." We *make* human equality as a matter of praxis. It is not something to be discovered in our objective natures or in Nature at large.

But if equality is a matter of practice, then it is also a matter of free choice. And in this case, there appears to be little separating her dim view of universal humanism and that of arch conservatives like Edmund Burke or Joseph de Maistre. The latter wrote in surprisingly similar terms: "But there is no such thing as man in the world. In my lifetime I have seen Frenchmen, Italians, Russians, etc.; thanks to Montesquieu, I even know that *one can be Persian*. But as for *man*, I declare I have never in my life met him."[84]

One must not lose sight of the deep intellectual threads which connect these seemingly polar-opposite worldviews. This consists in their common rejection of modernist reductionism, where diverse peoples, cultures, and places can be brought under the same, intelligible rubric. It consists, as well, in the suspicion of an allegedly cold, bloodless Science which has displaced traditional and local practices. In short, this is the resentment of a disenchanted world, and the desire to reintroduce a sense of novelty, vitality, and mystery.

Animism

The watchword for this reintroduction of spirit to the world is "animism." For the Gaian, existence is irreducibly alive. To speak metaphysically, pure Potentiality has preeminence over stable Being. The world is risky, noninnocent, and fluid. "Truly nothing is sterile; and that reality is a terrific danger, basic fact of life, and critter-making opportunity."[85]

---

83. Stengers, *In Catastrophic Times*, 136. There is a tendency among some contemporary authors to affirm a novel definition of "universals" or "universalism" while denying their transhistorical or transcultural nature. Again, however, our aim in this chapter is to focus on the content of a theorist's work rather than their chosen signifiers. In this, it is important to state plainly that "universals" which depend upon particular circumstances or contexts are a misnomer.
84. Joseph Maistre, *Considerations on France*, ed. and trans. Richard A. Lebrun (Cambridge: Cambridge University Press, 2006), 53.
85. Haraway, *Staying with the Trouble*, 64.

As intimated earlier, "animism" is derived from the claim that effects have a power and efficacy all their own, beyond their causal origins. Latour, for example, will speak of scientific theories as beginning with an observed "list of actions" before any determinate characterization of the same.[86] We perceive, manipulate, and make-use-of, before we understand, define, or taxonomize. Technology grounds science; practice grounds theory.

But the immediate result of such functionalism is the conferral of personality, agency, or in other words "life," to what are usually considered inanimate objects. "Having a function is its [i.e., a speculative object's] way of having goals, or in any case of being defined as a vector, and thus as an agent."[87] Thus, there is a conceptual collapse of the distinction between merely "doing a thing" and having specific agency or intentionality.

The eco-pessimist will sometimes defend this eccentric move by painting their "normal science" critics as entirely parasitic upon an animist worldview. It is pointed out that Newton, for example, believed in spooky "action at a distance," and that this was explained through recourse to angels (however superficially secularized):

> Through several hundred pages of angelology, Newton gradually managed to trim their wings and transform this new agent into a "force. [...] but it was still charged, upstream, by millennia of meditations on an "angelic system of instant messaging." As we know quite well, purity would sterilize the sciences: behind the force, the wings of angels are always beating invisibly.[88]

It is not as if the Gaians see themselves as illicitly animating what is truly dead. This is their semantic distinction between "animism" and "vitalism." Both positions reject the mechanist viewpoint which sees most of reality as inert and thus "overanimates" only certain, privileged beings, especially *Homo sapiens* (as opposed to their background environment). The vitalist, on the other hand, *too hastily* animates aspects of our environment, positing an ineffable "force" which is the author of all life. The animist, rejecting both of these "extremes," instead recognizes the animate, intentional qualities of each thing. Here, there is no need to posit any singular Lifeforce, Ether, or miasma. The "black box" noumenalism of their worldview rejects any such mysterious force; it rather insists upon a network of agencies which can be discovered, analyzed, and comprehended (if only in temporary and local ways).

---

86. Latour, *Facing Gaia*, 57.
87. Latour, *Facing Gaia*, 55.
88. Latour, *Facing Gaia*, 66.

This, for Latour, correlates to the nineteenth-century debates around fermentation. In this context, the chemist Liebig posited a wholly mechanistic explanation for the phenomenon, while Félix Pouchet instead sought to pin the whole process of fermentation upon a generic "heterogenesis" (where living things emerged, literally, from thin air). Pasteur, in contrast to both, argued that fermentation was indeed a living process, but that this had to be explained through particular biological processes, instead of some generic, vitalistic power.[89] Hence, the animist (Pasteur) is distinguished from both the mechanist (Liebig) and the vitalist (Pouchet) because of his specificity.

The crucial point is that animism is no mere *addition* of life, but instead, the sensible recognition of what is *already* alive. Rhetorically, the Gaian will accuse their opponents of performing an illegitimate reversal. The modern scientist will take a clearly observable, living world, and perversely pretend (after the fact) that it is nothing but a string of dead causes and effects.

> The idea of a deanimated world is only a way of linking animations *as if* nothing were happening there. But agency is always there, whatever we may do [...] it is a *secondary stylistic effect*, posterior, derived, through which we purport to *simplify* the distribution of actors by proceeding to designate some as animate and others as inanimate.[90]
>
> Animation is the essential phenomenon; and deanimation is the superficial, auxiliary, polemical, and often defensive phenomenon. One of the great enigmas of Western history is not that "there are still people naïve enough to believe in animism," but that many people still hold the rather naïve belief in a supposedly deanimated "material world."[91]

As with all things in the Gaian worldview, this theoretical notion of animism has a particular origin story and context. It may be that the world was always alive; nonetheless, our increasing recognition of this fact is timely. Global climate change, especially global warming, sea level rise, and increasingly severe weather, has forced the issue upon our collective consciousness. If the environment was once perceived to be an inert, passive "background," it is now the most significant, often violent, "actor" determining the course of our lives. The smug, modern sensibility that we have moved past the fear and trembling before Nature has been irrevocably lost.

---

89. Latour, *Facing Gaia*, 90.
90. Latour, *Facing Gaia*, 68.
91. Latour, *Facing Gaia*, 70.

Such is the meaning of the New Climate Regime: the "warming" is such that the old distance between background and foreground has faded away: it is *human* history that appears cold and *natural* history that is taking on a frenzied aspect. The metamorphic zone has become our common place: it is as though we had indeed ceased to be modern, and, this time, collectively.[92]

The trauma of climate change has reintroduced the premodern idea that activity and animation are to be found on both sides of the human/nonhuman divide.[93] For Haraway, this means a renewed respect for matter over form. If the logocentrism of Western philosophy debased matter as merely receptive, then Haraway insists that we recognize its indispensable activity. The soil is a prime example of this, where no plant life could thrive in a truly sterile, inert environment.

Planting seeds requires medium, soil, matter, mutter, mother [...] matter is never "mere" medium to the "informing" seed; rather, mixed in terra's carrier bag, kin and get have a much richer congress for worlding. Matter is a powerful, mindfully bodied word, the matrix and generatrix of things, kin to the riverine generatrix Oya. It doesn't take much digging or swimming to get to matter as source, ground, flux, reason, and consequential stuff—the matter of the thing.[94]

But animism is not only derived from the negation of the being/environment dichotomy. It also questions the modernist placing of primary qualities before secondary ones. As discussed earlier, physical extension is typically considered to be logically prior to an entity's ability to move, breathe, feed, or procreate. Yet, for the Gaian, this prioritization is just another illicit attempt at demoting the real qualities of life. The modern imagines that perceived, vital qualities are merely parasitic upon dead, mechanical relations and extensions.

This goes to the indelibly phenomenological approach of the eco-pessimists. We experience a world full of secondary qualities, that is, a world with colors, sounds, and scents, not to mention chemical, atmospheric, geological, and biological processes. To imagine a grey, passive landscape containing only sterile billiard balls that can be moved from without is really nothing more than an odd thought experiment.

---

92. Latour, *Facing Gaia*, 74.
93. Latour, *Down to Earth*, 76.
94. Haraway, *Staying with the Trouble*, 120.

At most, this was at one time a useful heuristic when humanity was trying to discover something about fundamental physics or the motions of heavenly bodies. But, according to the Gaians, this heuristic became outsized and then considered (wrongly) to be the way things truly are. "What was only a convenient expedient for Galileo was transformed into a metaphysical foundation in the hands of Locke, Descartes, and their successors."[95]

Lest readers worry that this animism will overturn science in favor of a bizarre, supernatural landscape, the eco-pessimists reassure us that they are the most sober of naturalists and materialists. As Haraway put it, referring to Viveiros, "Animism is the only sensible version of materialism."[96]

Their method for arguing this point is intriguing. Instead of denying spontaneous vital powers, they hope to secure naturalism by radically amplifying these very potencies. The idea is that if there are sufficiently diverse intentionalities, then these will effectively cancel each other out (or, at least, not allow any one to have a disproportionate agency over all the others). "The more you extend the notion of intentionality to all the actors, the less intentionality you will detect in the whole."[97]

The formula here is an echo of Deleuze and Guattari's dictum that "pluralism = monism." For the Gaian followers of Deleuze, these terms are substituted for "Pan-psychism = Materialism."[98] In any case, the strategy—ironically reminiscent of accelerationist theory—is to overburden one term until it turns over into its apparent opposite. Thus, an amped-up pan-psychism becomes a materialism on the other end.

As we can now see, Gaian "naturalism" is entirely distinct from the classic naturalism of Spinoza or the French materialists. In his *Système de la Nature*, d'Holbach states that human intentions, indecisions, and emotions are the result of a determinate "mechanism" relative to other bodies; similarly, Spinoza argues that humans are no special "kingdom within a kingdom," but subject to the very same physical laws as all other entities.[99]

In stark contrast, the Gaian tries to avoid anthropocentrism by "levelling up" the agency of the surrounding environment. It is not that we are just as deterministic as the boulder; instead, the rocks are just as alive as we are.[100]

---

95. Latour, *Facing Gaia*, 85.
96. Haraway, *Staying with the Trouble*, 88.
97. Latour, *Facing Gaia*, 99.
98. Danowski and Viveiros de Castro, *The Ends of the World*, 113.
99. "I shall consider human actions and appetites just as if it were a Question of lines, planes, and bodies." Spinoza, "Ethics," E3 Preface; Paul-Henri Thiry Baron d'Holbach, *The System of Nature*, Volume 1 (Whitefish, MT: Kessinger Publishing, 2004), 136.
100. Or if not "just as" alive, then at least on the same animistic continuum of life.

And when Western philosophy penitentially undergoes self-critique and tries to attack anthropocentrism, its way of negating human exceptionalism consists in affirming that we are, at a fundamental level, animals, or living beings, or material systems *like all the rest*; "materialist" reduction or elimination is the favored method for bringing humans down to the same level as the pre-existing world [...] [I]n contrast [...] "pan-psychic" generalization or expansion is the basic method for bringing the world up to the same level [as human beings].[101]

This is a problematic strategy for achieving naturalism. Merely cancelling out a single, strong will with a chorus of diverse wills does not avoid vitalism. Either one affirms a universe which is governed by intelligible, physical laws (classic materialism), or one rejects this, and instead affirms a reality which is spontaneously self-moving. Laundering self-movement and the free will through a plurality of agents does nothing to naturalize this picture. Natural laws are either "lawful," or they are not worthy of the name to begin with.

The reason why natural laws, in their modern sense, must be rejected by the Gaian is that they are impersonal. But given the phenomenological approach of these eco-pessimist thinkers, nothing can be wholly impersonal and objective. Instead, they affirm a species of "correlationism." This consists of the idea that material reality is inherently bound up in thought, or, put negatively, that "unthinking Being" is impossible.

The Gaian is therefore opposed to much of accelerationist epistemology, itself influenced by the speculative realist movement. For the accelerationist, it is important to try and gain insight into a pure, inhuman Being, for example, through mathematical intuitions.[102] This is all part of their strident anti-humanism, and the desire to escape from all parochial, biologically imposed limits.

---

101. Danowski and Viveiros de Castro, *The Ends of the World*, 72.
102. Danowski and Viveiros are here criticizing the works of Quentin Meillassoux and Ray Brassier, who in their estimation envision a world which is "radically dead." As they put it, "matter, if it is to exist in itself (outside correlation), must be passive and inert—in the sense of insentient, indifferent, and meaningless—reintroduces the human exceptionalism that it purported to eliminate. The anti-anthropocentric decision at the root of these two versions of the 'world without us' theme reveals itself to be, when all is said and done, obsessed with the human point of view. It is as if the negation of this point of view were a necessary condition for the world to exist—a curious negative idealism, a weird cadaverous subjectalism [...] But a negative anthropocentrism is still an anthropocentrism—perhaps the only really radical one." Danowski and Viveiros de Castro, *The Ends of the World*, 31, 35.

For the Gaian, on the other hand, there is no radical outside or "Great Outdoors" to thinking. There is only, we may say, the "Great Indoors." They embrace the fact that one cannot escape the intentional, agentic ways in which we perceive the world. "Worlds" are created through a multiagent process of "worlding." Even the "end of the world," opines Danowski and Viveiros, can only be thought in a "correlationist" manner, that is, as a "problem posed by thought, since only thought problematizes."[103]

This correlationism, nonetheless, does nothing to mitigate the Gaian's exuberant animism. To the contrary, their thinking appears to be that if *all* agents perceive the world, then it is not only human consciousness which correlates with reality. "If one de-privatizes correlationism, one arrives quickly at some idea that everything has agency, everything is 'alive,' possibly 'conscious;' or that consciousness is just another mode of access among equal others, and so on."[104] Thus, we have here the attempt at a nonanthropocentric worldview which, nonetheless, is inextricably tied to consciousness.

For Danowski and Viveiros, in particular, this worldview is expressed as an "anthropomorphism" but never an "anthropocentrism." While the difference in these terms may appear negligible to a lay person, for these thinkers it is an indispensable distinction. Anthropocentrism is the bad, modernist idea that human beings are more important (whether morally, metaphysically, or epistemologically) than other creatures.

Anthropo*morphism*, by contrast, is the idea that all things are of one body. The entire world is inseparable from the human form, and conversely, the human form is of one flesh with the world. Danowski and Viveiros take inspiration from particular Amerindian creation myths which feature a primordial "human" who, through folly, engenders the rest of existence.[105] Hence their correlationism: "There may have been a humanity before the world; but there can be no world after humanity, that is, a world that lacks relation and otherness."[106]

Again, this animism and anthropomorphism give no special import or authority to *Homo sapiens*. It is not as if our species is the locus of meaning and definition for every other being (as with Genesis 2 and Adam's naming of the animals). We do not "crown the Great Chain of Being by adding a new ontological layer."[107] Indeed, this strong correlationism appears to have the reverse effect. If the world (i.e., things-in-themselves) is so bound up with the human

---

103. Danowski and Viveiros de Castro, *The Ends of the World*, 20.
104. Morton, *Humankind*, 51.
105. Danowski and Viveiros de Castro, *The Ends of the World*, 63.
106. Danowski and Viveiros de Castro, *The Ends of the World*, 78.
107. Danowski and Viveiros de Castro, *The Ends of the World*, 65.

form, then biconditionally, the human form becomes just as mysterious and anomalous as the rest of creation.

Human exceptionalism is, ironically, negated by the very fact that the world is of our body. "To say that everything is human is to say that humans are not a special species."[108] In this scheme, we are neither in possession of a transparent, intelligible "human nature" nor does our essence determine the natures of other living things. Rather, it is our own, inherent plurality which gave rise to the plurality of the universe.

> Things did not wait for a human arche-namer to get to know *that* they were, and *what* they were. Everything was *human*, but everything was not *one*. Humankind was a polynomic multitude; it appeared from the start in the form of an internal multiplicity whose morphological externalization—that is, speciation—is precisely the stuff of cosmogonic narrative.[109]

Finally, we arrive at the formulation of Gaian anti-humanism (though they would likely balk at the term). The conceptual movement here is one from epistemology to ethics. In the first instance, the human being becomes anomalous and unknowable. "When everything is human, the human becomes a wholly other thing."[110] In Haraway's words, "bounded individualism [...] has finally become unavailable to think with, truly no longer thinkable."[111] In fact, there can be no bounded individuals (human or otherwise), since all species co-compose their worlds. Hence, Haraway prefers the term "humus" (connoting the complex mélange of fertile soil) to *Homo*.[112]

This rejection of the individual has subsequent effects for how one thinks about politics, economics, and psychology. Specifically, Freudian and other ego-oriented psychologies have to be thrown into question. This is very likely Haraway's target when she takes aim at "Western-influenced psychology [...] besotted by individualism."[113] So do political and economic theories (neoliberal or otherwise) which emphasize individual rights, responsibilities, and welfare. Any theory which envisions humans as distinct from their environments, that is, as "bounded individuals plus contexts," is already off to a bad start.[114]

---

108. Danowski and Viveiros de Castro, *The Ends of the World*, 72.
109. Danowski and Viveiros de Castro, *The Ends of the World*, 67.
110. Viveiros de Castro, *Cannibal Metaphysics*, 63.
111. Haraway, *Staying with the Trouble*, 5.
112. Haraway, *Staying with the Trouble*, 55.
113. Haraway, *Staying with the Trouble*, 18, 19.
114. Haraway, *Staying with the Trouble*, 30.

This descriptive rejection of the individual then takes on a strongly normative tone, replete with suggestions for politics. In *Staying with the Trouble*, Haraway is keen to repeat her slogan, "Make Kin Not Babies!"[115] Couched in feminist discourse of transcending the traditional nuclear family, her ultimate prescription is one of gradual (albeit massive) depopulation.

While not all efforts at population control are necessarily anti-humanist (contraception has certainly been an empowering force for humans), there is something distinctly Malthusian about Haraway's prescriptions. It is as if having greater numbers of people is not only contingently bad, because of Earth's carrying capacity, but that this is a perverse aim in itself. The desire to increase and improve humanity's estate is diagnosed as an obsession with "progress."

For their part, Danowski and Viveiros explicitly invoke Malthus in promoting an Amerindian ethos of sufficiency as opposed to "Western" notions of unlimited progress. They see this "slow" model of society as prefiguring the only way forward for humanity as it faces climate change and resource depletion.

> Ultimately, Indians prefer to maintain a relatively stable population instead of increasing "productivity" and "improving" technology in order to create conditions ("surplus") so that there can always be more people, more needs, more concerns. The ethnographic present of slow societies contains an image of their future. Indians are Malthusians in their own way.[116]

Dipesh Chakrabarty similarly rejects humanistic ends for grounding international public policy. Even the notion of "sustainability," he claims, is too bound up with the "humancentric" idea of "maximum sustainable yield."[117] But Chakrabarty abhors the notion that we should treat nature as a resource to begin with. "Scientific management" of forests and fisheries merely presumes that the desired end is the promotion, increase, and welfare of the human species. This, for Chakrabarty, illegitimately de-emphasizes the inherent value of the rest of nature. "In the literature on sustainability, earthly processes constitute a mute background for human activities."[118] Rather than focus

---

115. Haraway, *Staying with the Trouble*, 102.
116. Danowski and Viveiros de Castro, *The Ends of the World*, 104.
117. Dipesh Chakrabarty, "The Planet: An Emergent Humanist Category," *Critical Inquiry* 46, no. 1 (Autumn 2019): 19.
118. Chakrabarty, "The Planet," 20.

on human flourishing (or even the welfare of sentient species), Chakrabarty promotes an explicit biocentrism. *Life itself* is the locus of moral concern.

> The key term in planetary thinking that one could contrapose to the idea of *sustainability* in global thought is habitability. Habitability does not reference humans. Its central concern is life, complex, multicellular life, in general, and what makes *that*, not humans alone, sustainable.[119]

One may rightly ask, given such biocentrism, how ethical choices can be made? This certainly seems to be a meaningful question for the Gaian thinkers. While Haraway will insist that we "make kin, not babies," she immediately follows up with the warning that "it *matters* how kin generate kin."[120] In her documentary *Story Telling for Earthly Survival* (2016), she insists, "We are for some worlds/ways of life and not others."[121]

But how can we even begin to order our ethical priorities when what matters is "life itself" (irrespective of a being's capacity to think or feel)? How can we, for example, decide between treating a dog for fleas and catering to the fleas themselves?

While such a basic question may appear impertinent, it is posed with the utmost sincerity. For the sort of anti-humanism promoted by the eco-pessimist is not merely the rejection of *speciesism* (i.e., the superstition that humans matter more than other species "just because"). It is instead the rejection of *sentience* as the major criterion for moral consideration. Their strong denial of individualism, of any separation of individual and environment, entails this radical position.

Consistent with their ethics, Gaian policy prescriptions likewise depart from a humanistic Marxism. Stengers chides "Marx's inheritors" with "fabricating a point of view organized around a humanist version of salvation." This bad form of Marxism is supposedly obsessed with a universal conception of human nature. Here, humanity's world-historical goal is to finally overcome "what separates it from its truth."[122] For the Gaian, as we have seen, there can be no stable, transhistorical "truth" of human nature to achieve. Stengers goes

---

119. Chakrabarty, "The Planet," 20.
120. Haraway, *Staying with the Trouble*, 103; emphasis is ours.
121. Terranova, *Story Telling for Earthly Survival*.
122. Stengers, *In Catastrophic Times*, 151–52. This humanism is perhaps best seen in Marx's discussion of species-being in the Paris Manuscripts of 1844, specifically in the section on Estranged Labor. See Karl Marx and Frederick Engels, *Economic and Philosophical Manuscripts of 1844 and the Communist Manifesto*, trans. Martin Milligan (Amherst, NY: Prometheus, 1988), 75–78.

so far as to express her "dread" at "appeals to sacred unity" which inevitably attend such essentialist projects.[123]

Gaian animism is constitutionally opposed to universal human nature because it implies a static, totalizing essence. Not only this, but "human nature" implies an originary "Nature" to which it is traditionally opposed (as in the nature/culture division). But for Gaian metaphysics, there is no distinction between human flesh and the matter of the world. All things are alive, and to posit an inert "background" to an active humanity is not only empirically wrong but also morally violent. Violence also marks all efforts at final, comprehensive knowledge, rather than a respect for the constant co-worlding of diverse "critters."

Humility

The prime virtue of the eco-pessimist is humility. However, these authors do not understand humility as a resignation of all activity. One must be resolutely political and committed to struggle. But for all that, there can be no final victories and no final reconciliation with our "true" essence. In Clive Hamilton's writing, this is linked to a Heideggerian critique of Enlightenment morality, especially the notion of progress and moral perfectionism. As he put it, "One thing is now clear; the perfectibility of humankind is a failed project."[124]

Stengers makes a very similar point, but now with reference to the perceived humility of premodern, traditional peoples. "Struggle there must be, but it doesn't have, can no longer have, the advent of a humanity finally liberated from all transcendence as its aim. *We will always have to reckon with Gaia*, to learn, like peoples of old, not to offend her."[125]

Gaia is the "great power of *deflation*" which counters the modernist's presumption of their own superiority, as "the frog swollen with air who believes that it is bigger than an ox."[126] Through catastrophic climate change, Gaia enforces a regime of humility upon us. This is a blow to modern hubris and its "epic" notion of civilizational history.

This Gaian dispensation has very practical implications for how we should live our lives. Consumerism, especially the idea that the next generation should

---

123. Stengers, *In Catastrophic Times*, 56.
124. Hamilton, "When Earth Juts Through World." See Hamilton's discussion of the Davos conference debate between Heidegger and the neo Kantian Ernst Cassirer.
125. Stengers, *In Catastrophic Times*, 58. Haraway makes an analogous point on the need for only "partial recuperation" and "getting on together" as opposed to "reconciliation or restoration." Haraway, *Staying with the Trouble*, 10.
126. Latour, *Facing Gaia*, 289.

have more comforts and amenities than the last, is a sickness. Danowski and Viveiros contrast this "American dream" of endless material progress with the Andean motto: "*vivir bien, no mejor* (to live well, not better)."[127] Put otherwise, one should aim at sufficiency instead of improvement.

But this chastening of our acquisitive spirits is hard for the modern to accept. Its advocates are eyed with suspicion and painted as agents of regression:

> Just as we once abhorred the vacuum, today we find repugnant the very idea of deceleration, regression, retreat, limitation, degrowth, applying brakes, descent—*sufficiency*. Anything that brings to mind any of these movements toward an intensive sufficiency of world (instead of an epic overcoming of the "limits" separating us from a hyperworld) is immediately accused of naive localism, primitivism, irrationalism, bad conscience, guilt, or even just fascistic tendencies, period.[128]

If some notion of progress is retained by the Gaians, then it is only one of "non-material intensification." We should aim for an improvement in our quality of life, especially when it comes to leisure, rootedness to the land, and the revival of our customs and traditions. These aims are contrasted to "Promethean mastery," "managerial control" and other quantitative approaches to progress.[129]

Certainly, it would be a mistake to equate human happiness with mere quantitative increases in personal wealth or GDP. There are, undoubtedly, diminishing returns when it comes to the amount of happiness each additional dollar can buy. (If we permit ourselves to speak of "buying happiness" to begin with) Nonetheless, it is hard to deny the simple, materialist proposition that human flourishing does require a base-level of material comfort, and that higher orders of culture, education, and even entertainment can often be achieved only with the aid of greater resources. The World Happiness Report strongly confirms this supposition, where the happiest five countries are all wealthy nations with strong social safety nets, infrastructure, and education systems.[130]

---

127. Danowski and Viveiros de Castro, *The Ends of the World*, 76.
128. Danowski and Viveiros de Castro, *The Ends of the World*, 21.
129. Danowski and Viveiros de Castro, *The Ends of the World*, 97.
130. John F. Helliwell et al., eds., "World Happiness Report 2020" (Sustainable Development Solutions Network, n.d.), https://worldhappiness.report/ed/2020/. World happiness, at least according to this and similar surveys, does not seem to directly correlate to personal income alone. Nonetheless, social wealth (e.g., in the form of a well-functioning national health system) is still a form of material wealth that clearly seems to correlate with human happiness.

In any case, there is a difference between a slavish dependence on luxury (think the status-obsessed Patrick Bateman in *American Psycho*), and the ability to make use of nonessential, material goods. The latter, we suppose, is actually a virtue. As Seneca affirmed of the wise man, he "lacked nothing but needed a great number of things."[131] In other words, it is a sign of human flourishing that one can appreciate a well-prepared meal, a night at the theater, or a decent painting set. One needn't feel ashamed of having good taste and enjoying oneself. "Nothing forbids our pleasure except a savage and sad superstition."[132] If certain patterns of consumption carry with them outsized environmental consequences, then certainly this should give us pause. Nonetheless, this is a question of balancing two kinds of goods (personal cultivation and environmental sustainability), rather than denigrating the former as a "sickness."

Beyond a mere critique of the modern lifestyle, the virtue of humility is seen as transforming the domains of science, politics, and religion. For Latour, the Ash Wednesday phrase "Remember that you are dust and to dust you will return!" is not a curse but a blessing. This imperative to humble ourselves is transformative of our intellectual, social, and spiritual endeavors.[133]

Gaian humility also has far-reaching consequences for the critique of big Science. The "intrusion of Gaia," in the form of climate catastrophe, has supposedly shown that this discipline is not "equipped to respond to the threats of the future" if we hope to avoid an impending barbarism.[134] Specifically, the Gaian rejects all a priori limits frequently placed on "normal" science. For if we are epistemically humble, then we cannot presume ahead of time that all of observable reality will conform to our intellectualist demands (especially that of an orderly, "clockwork" universe). Humility likewise requires that we be open to observing the inherently animate, agentic qualities of the world—not mistaking anything as inert, or as an instance of mere mechanism.

Moreover, a humble comportment toward science will remind us that this discipline does not exist in a vacuum. Here, the phenomenological approach of the Gaians is particularly manifest. Latour insists that science cannot be thought apart from anthropology, that is, a cognizance of the specific people doing the work in the lab and in the field. Ripping away the timeless, placeless veneer of modern science thereby deprives it of its infallibility. Science is made to be "modest," no longer staking out timeless truth claims, but merely acting in a humble, instrumental capacity.

---

131. Lucius Annaeus Seneca, *Letters from a Stoic*, trans. Robin Campbell (London: Penguin Classics, 1969), Letter IX. Seneca is here quoting Chrysippus.
132. Spinoza, "Ethics," E4P45 Schol.
133. Latour, *Facing Gaia*, 285–86.
134. Stengers, *In Catastrophic Times*, 29.

And yet it [Gaia] metamorphizes the sciences effectively and will change them forever: it anthropologizes them, brings them back down to Earth, encourages their multiplicity, welcomes their instrumentation, conspires with their rediscovered modesty. Gaia requires the sciences to say where they are situated and what portion of Earth they inhabit.[135]

Part of this humiliation involves taking away science's status as neutral, "unchallenged arbiter" in otherwise social conflicts. Latour, especially, makes the point that it is the alleged neutrality of the sciences which have traditionally given them outsized normative power.[136] After all, who are we to argue with the objective facts?

But for the eco-pessimist, there are no "mere" facts. Everything is alive and bound up with consciousness. And because consciousness is always particular and moving, everything is likewise part of an intentional agenda. "*Matters of fact* [...] have become so many *matters of concern*."[137] Neutrality is understood to be a perverse myth. Thus, when the scientist claims special authority by appearing totally dispassionate, this can be nothing more than a deceit.

Every individual has an agenda, a perspective, and a point of view. "Speaking in a mechanical voice is still speaking."[138] If science is to reform itself, then it should begin with the recognition of its nonneutrality; the "earthbound scientist," according to the Gaian, should quit pretending to be a third party, and instead admit that they are "just one party" within a grand contest of wills and agendas.[139]

If there are no neutral facts, and no a priori guardrails to what can be hypothesized, then the possible objects of science become limitless.[140] This opens the gates to all manner of supernatural and occult speculations. Stengers, for example, seems to elide the difference between rural healers "without formal qualifications" and the New Age spiritualism of a Franz Mesmer.[141]

For his part, Morton praises Mesmer (along with the mystic Emanuel Swedenborg) for the theory of animal magnetism, that is, "a force that surrounds and penetrates lifeforms and acts in a nonlocal, telekinetic, telepathic

---

135. Latour, *Facing Gaia*, 289.
136. Latour, *Facing Gaia*, 23.
137. Latour, *Facing Gaia*, 164.
138. Latour, *Facing Gaia*, 68.
139. Latour, *Facing Gaia*, 253.
140. Latour, and others, will sometimes speak of the need for limits to the empirical sciences. However, given their rejection of the a priori, it is unclear what such limits could amount to. See Latour, *Down to Earth*, 78.
141. Stengers, *In Catastrophic Times*, 129.

and hypnotic way."[142] He only criticizes the latter-day appropriations of this theory, which transform such a "nonlocal" force into a personal superpower. Stengers elsewhere laments the modern dismissal of "seers," "Tarot readers," and "cowrie shell diviners" as mere instances of superstition.[143] But a properly humbled science, claim Danowski and Viveiros, will reacquaint itself with traditional folklore, from which it originally separated three thousand years ago.[144]

This transformation of science is indicative of a deeper postmodernism. A contested term (few these days accept the label), it at least denotes the rejection of any transcendental signified. Put more simply, the postmodernist rejects the existence of any objective realities apart from some kind of discourse or agenda. In her earlier writings on Cyborg theory, Haraway likewise claims that "the transcendent authorization of interpretation is lost, and with it the ontology grounding 'Western' epistemology."[145]

Far from rejecting Haraway's early position, the eco-pessimists appear to have elevated it to a sort of "negative theology." But unlike the negative theology of a Maimonides or Cusanus, this apophatism does not gesture toward an unknowable infinite. For Gaia is always composed, plural, and finite. "This *via negativa* is as injunction to humility that not only refuses representation but also the indexical and the holistic."[146]

The normative corollary to such apophatism is a species of "care ethics." It is a focus on dependency and fragility as opposed to autonomy and independence. As Haraway claims, "Nobody lives everywhere; everybody lives somewhere. Nothing is connected to everything; everything is connected to something."[147] What matters are not timeless, universal theories of "the good," but the ability to tell compelling stories about specific situations. Such stories, Haraway surmises, can make us more receptive to particular instances of joy and suffering, and so allow us to become more "response-able" before the other. General ethical theories do not have preeminence over the particular situations to which they are applied; to the contrary, specific situations are of the greatest consequence. "The details matter. The details link actual beings to actual response-abilities."[148]

---

142. Morton, *Humankind*, 50–51.
143. Stengers, *In Catastrophic Times*, 149.
144. Danowski and Viveiros de Castro, *The Ends of the World*, 6.
145. Haraway, *A Cyborg Manifesto*, 12.
146. Wark, *General Intellects*, 314.
147. Haraway, *Staying with the Trouble*, 31.
148. Haraway, *Staying with the Trouble*, 115, 29.

Haraway makes this clear in seeming to defend the practice of pigeon racing (as against the objections of PETA). For the specific cultural and artistic enjoyment of this practice may excuse some of the physical harm to the birds themselves. The rich historical narrative of this practice is just as significant as, for instance, medical breakthroughs derived from other instances of animal exploitation.[149] There can be no universal ethical theory (utilitarian or otherwise) which can decide such controversies ahead of time. Storytelling, and specific contexts, will always have a role to play.

This care-ethical approach, with its focus on concrete interdependencies, thus implies a blurring of the distinction between ethical agents and patients. Since everything is alive, and a responsible actor, nothing at all is entirely "innocent" or passive. Chakrabarty even counts industrialized farm animals as responsible agents. For the methane they produce has deleterious effects on climate change. This absolute diffusion of culpability is only possible given an animistic flattening of the moral universe, where mere "causality" is equated with autonomy and intentional agency. In this way, "causal responsibility remains distributed."[150]

Haraway also sanitizes the practice of mass animal slaughter through highly euphemistic language. Bizarrely, she characterizes the raising of animals for meat as a "mutual," collaborative activity. That humans "nourish" animals, and vice versa, obviously ignores the stark power differentials in the animal/slaughterer relationship. If factory-farmed veal could speak, would they truly express their condition as one of "working together" with the farmer?

> The result is bringing into being animals that nourish humans, and humans that nourish animals. Living and dying are both in play. "Working together" in this kind of daily interaction of labor, conversation, and attention seems to me to be the right idiom.[151]

On this view, there can be no moral panacea when it comes to our treatment of nonhuman animals. In fact, posing the question in such an anthropocentric way (i.e., "*our* treatment of animals") is already flawed, according to the Gaian. Rather, it is always a political question of living and working *along with* other political actors.

---

149. Haraway, *Staying with the Trouble*, 23–24.
150. Dipesh Chakrabarty, "The Human Condition in the Anthropocene" (The Tanner Lectures in Human Values, Yale University, 2015), 171, https://tannerlectures.utah.edu/.
151. Haraway, *Staying with the Trouble*, 129.

> We also avoid the trap of thinking that it would be possible to live in sympathy, in harmony, with the so-called "natural" agents. We are not seeking agreement among all these overlapping agents, but we are learning to be dependent on them. No reduction, no harmony. The list of actors simply grows longer; the actors' interests are encroaching on one another; all our powers of investigation are needed if we are to begin to find our place among these other actors.[152]

The main point here is that, if multispecies relations are irreducibly political and "noninnocent," then they are also always fragile. Latour is triumphalist on this point: "Finally, multiplicity is everywhere! Politics can begin again."[153] In other words, there is the constant threat of alliance breakdown, and even of war. There is no preset plan or immutable World Order which can dictate when peace reigns and when wars begin. It is always, irreducibly, a matter of *decision*. Freedom, politics, and autonomy are nothing if not matters of the will.[154]

If the will has priority over an objective, lawful, and intelligible universe, then the possibilities of thoroughly understanding the other are radically circumscribed. For each culture is, at least in part, a free creation of a particular people. They cannot be reduced to just so many modifications of a universal (and thus transparent) "human nature." This hesitance to understand the other is often cast as a respectful rejection of cultural appropriation.

> For we cannot think like Indians; at most, we can think with them. And on this point, (to attempt, but of course just for a moment, to think "like them"), it should be said that if there is a clear message in Amerindian perspectivism, it is that one should never try to actualize the world that is expressed in the gaze of the other.[155]

Likewise, Haraway will express her feeling that, as a "daughter of [White] conquest," she has no right to own a Navajo basket. (Though, of course, she won't give it up, as no one is entirely innocent anyway.)[156] She will consume

---

152. Latour, *Down to Earth*, 87.
153. Latour, *Facing Gaia*, 143. Latour sometimes hedges on this point: "Obviously there is no politics other than that of humans, and for their benefit! This has never been in question. The question has always been about the form and the composition of this human." See Latour, *Down to Earth*, 85.
154. Latour, *Facing Gaia*, 275.
155. Viveiros de Castro, *Cannibal Metaphysics*, 196.
156. Terranova, *Story Telling for Earthly Survival*.

indigenous cultural products (from ancient stories to video games), but nevertheless insists that "the conditions for sharing stories must not be set by raiders, academic or otherwise."[157] In this way, the Gaian wants to have their cake and eat it too: One can draw inspiration from other cultures, as long as we don't presume to understand them too much. We can own their artifacts, but only if we make the obligatory allusions to the whole "noninnocent" history of colonization and genocide.[158]

In fact, there appears to be a surprising moral boon which comes with inflicting harm on other beings. For this only reminds us of our inherently political and noninnocent relation to them. And, in keeping with the underlying care-ethical approach, these particular connections (even when harmful) make us more "accountable" and "response-able" to those very beings.

Haraway is explicit about this moral dynamic in her far-ranging discussion of Premarin and synthetic estrogen (DES), originally derived from the urine of female horses. The extraction of these compounds involves the farming, unnatural confinement, and eventual slaughter of mares. But, again, Haraway construes this trauma as a potential ethical benefit:

> Having eaten Premarin makes me more accountable to the well-being of ranchers, northern prairie ecologies, horses, activists, scientists, and women with breast cancer than I would otherwise be. Giving my dog DES makes me accountable to histories and ongoing possibilities differently than if we never shaped kinships with the attachment sites of this molecule[...] We are all responsible to and for shaping conditions for multispecies flourishing in the face of terrible histories, but not in the same ways. The differences matter—in ecologies, economies, species, lives.[159]

---

157. Haraway, *Staying with the Trouble*, 87.
158. In a late 2019 interview for the *Los Angeles Review of Books*, Haraway makes an analogous point about "imperialist" appropriation of animal consciousness. When asked if she would like to know exactly what her dog Cayenne is thinking, she expresses horror at the prospect, arguing that full knowledge is a "violent fantasy." Any fantasy of "perfect communication" leads to the prospect of "murder and war." Instead of knowing, we need to honor the otherness of beings with nonknowing. "If you take anybody seriously, one of the things you learn is not knowing. That's one thing I learned from Cayenne and my other dogs. Not knowing is a quasi-Buddhist value." Steve Paulson, "Making Kin: An Interview with Donna Haraway," *Los Angeles Review of Books*, December 9, 2019.
159. Haraway, *Staying with the Trouble*, 116.

But the question remains as to whether this is a sound moral argument. Do we really need to inflict harm on a species (or another culture) in order to recognize our moral responsibility toward them? Even if this were somehow necessary, would it be morally justified? One might make the intuitive argument that *not* slaughtering animals is more beneficial to them, as opposed to confining and slaughtering them (even if the latter garners some deeper, "multispecies" connection).

In the end, the Gaian position is truly religious. Haraway's chapter on Premarin, titled "Awash in Urine," is evocative of the Christian image, "awash in the blood of the lamb." The eco-pessimist might suggest that the images are in stark contrast to one another. After all, the lamb is the symbol of innocence, and we are forgiven our sinful natures through partaking in it. But the urine of the mare is the ultimate symbol of noninnocence. It only reminds us of our guilt and culpability. On the other hand, in a thoroughly flattened moral universe where everyone is an actor, and no one is innocent, any specific culpability appears to get washed out in the whole, pluralistic mélange of actions.

If the Gaian is less than convincing when it comes to moral discourse, this is perhaps because they do not hold on to any ethical doctrine as distinct from their politics. The pluralism of their worldview means that real distinctions are always timely and relative. We can only define ourselves as against a specific other. "Animism taken to its final conclusion—as only the Indians know how to do—is not only a perspectivism but an "enemyism.""[160]

Here, one immediately notices the outsized influence of the Nazi jurist Carl Schmitt upon eco-pessimist thinking. In particular, his notion of the friend/enemy distinction suffuses much of their arguments.[161] Schmitt's overarching thesis rejects any rationalist account of the world as the falsification of "the political." While Latour, for his part, recognizes the problematic biography of Schmitt, he treats the reactionary theorist as merely a "poison kept in a laboratory," where "it is all a matter of dosage." (This is strongly reminiscent of Stenger's use of the "pharmakon" motif.)[162] One needs a measure of agonistic thinking in order to overcome modern paralysis, whereby we passively give ourselves over to the rule of technocrats and experts.

---

160. Viveiros de Castro, *Cannibal Metaphysics*, 194.
161. Latour, *Facing Gaia*, 235.
162. Latour, *Facing Gaia*, 228. Haraway seems to reject the explicit use of Carl Schmitt, claiming that Latour's "Gaia stories deserve better companions in storytelling." Of course, as we have seen, Haraway's own emphasis on political noninnocence is, itself, reminiscent of Schmitt's theories where every political actor is either friend or enemy. Haraway, *Staying with the Trouble*, 43.

The so-called New Climate Regime (of global warming) has shocked a portion of humanity out of its slumber. Before, it was not possible to do real politics, since we could not recognize "real opponents," but only "irrational people or infidels" (i.e., individuals who rejected climate science).[163] But now, with the crisis at full pitch, global warming has become a genuinely political affair, tantamount to a total war with existential ramifications. Belief in science is no longer sufficient; one must choose sides in the coming battle.

While it would be absurd to deny the serious menace of climate change, it is also crucial to note the conceptual sleight of hand at work here: The eco-pessimist pretends that "choosing sides" is a free and political decision, given how serious the threats are. But our knowledge of the existential threat is only possible because of the objective, quantifiable work of climate scientists. It is, in fact, *not* a free will decision what side of the climate wars one finds oneself; it is, rather, a matter of evidence and scientific literacy after all.

In any case, for the eco-pessimist, there is something inherently violent about Reason: To claim knowledge about objective moral standards can only lead to total wars of "limitless extermination."[164] Once we stop seeing our opponents as enemies, and begin to see them as outlaws, criminals, and moral reprobates, then there is no end to the sort of coercive, violent penalties we may visit upon them. "Globalist" claims about the objectively *right* sort of morality or science are just as violent as universalist claims about the *right* sort of religion:

> Let us recall Schmitt's argument: wars waged in the name of reason, morality, and calculations—the "just" wars—are the ones that lead to limitless extermination. Global wars waged in the name of the survival of the Globe would be much worse than the ones called "world wars." The extent, the duration, and the intensity of such wars can be *limited* only if we agree that the composition of the common world has *not yet been achieved*, that there is no Globe. How can we decide on these limits? By accepting finitude: that of politics and of the sciences, but also of religions.[165]

---

163. Latour, *Facing Gaia*, 223.
164. Latour, *Facing Gaia*, 285.
165. Latour, *Facing Gaia*, 285. Haraway echoes the same point in stating that "those who operate within the categories of Authority and of belief are notoriously prone to exterminationist and genocidal combat (it's hard to deny that!)." See Haraway, *Staying with the Trouble*, 42.

Of course, what Latour conveniently ignores here is that most paradigmatic example of total, exterminationist warfare in the twentieth century, that is, the Nazi campaigns for global domination. Schmitt himself provided legal justifications for the Nazi project to annihilate the Soviet Union, since the ideology of Bolshevism was internationalist, and thus undermined the political "pluriverse" of Europe. In Schmitt's *Nomos of the Earth* (written in the early 1940s), it was fascism, and not Bolshevism, that recognized finitude and "concrete-order thinking."[166] Schmitt also targeted Jewish consciousness (especially the assimilated Jew) as particularly dangerous given their supposed infidelity to any particular land or country.[167]

Nonetheless, Latour seems sanguine in flirting with highly reactionary jargon (even as he often professes liberal, internationalist policy prescriptions). In countering the globalization and alleged depoliticization of world discourse, he admits, "in practice, we are all counter-revolutionaries."[168] Danowski and Viveiros, for their part, offer something of an apologia for the concept of "ethnocentrism." While not good in itself, it is inherent to the (Amerindian) perspectivism which they celebrate. Since everything is potentially a human being, then conversely, Amerindians are liable to "deny humanity to their fellow men [...] even (or above all) to their closest geographical or historical cousins."[169]

While none of these theorists explicitly endorses ethnocentric policies, it is hard to see what conceptual resources they have to oppose such chauvinism. Lacking any universal notion of ethics, or a universal conception of humanity, all that remains is the competition of wills, represented as distinct races, ethnicities, and even species.

At most, there appears to be a superficial attempt to distinguish between "good" ethnocentrism and "bad" ethnocentrism. Latour, for example, counterposes the "*Blut und Boden*" (blood and soil) rhetoric of the Nazis with his preferred jargon of the "good old Earth" or the "earthbound."[170] Regardless of word choice, however, both phrases involve the rejection of an epic sort of emancipation, and an affirmation of attachment to the soil.

---

166. Domenico Losurdo, *War and Revolution: Rethinking the Twentieth Century*, trans. Gregory Elliot (London: Verso, n.d.), 107; Raphael Gross, *Carl Schmitt and the Jews: The "Jewish Question," the Holocaust, and German Legal Theory* (Madison: University of Wisconsin Press, 2007).
167. Mika Ojakangas, "Carl Schmitt's Real Enemy: The Citizen of the Non-Exclusive Democratic Community?," *European Legacy* 8, no. 4 (2003): 411 24.
168. Latour, *Facing Gaia*, 40. For examples of Latour's professed liberal internationalist stances, see Latour, *Down to Earth*, 100–106.
169. Viveiros de Castro, *Cannibal Metaphysics*, 58.
170. Latour, *Facing Gaia*, 243–44.

Politically, this notion of attachment always attends an ethos of traditionalism and antiglobalism. "Is it possible to make those who are still enthusiastic about globalization understand that it is normal, that it is just, that it is indispensable to want to preserve, maintain, ensure one's belonging to a land, a place, a soil, a community, a space, a milieu, a way of life, a trade, a skill?"[171] But such localism has the ultimate effect of dividing up, not only regions and territories, but peoples as well.

In line with Danowski and Viveiros's discussion of ethnocentrism, Latour insists that any political ecology must begin by "acknowledging the *division* of a human species that has been prematurely unified."[172] In other words, to posit a universal "human nature" is not only a metaphysical error, but a political one as well. The diverse "people of Gaia" are still plural, and potentially in deadly conflict with one another. They may come together and assemble, but this is far from a foregone conclusion.

While this pluriverse of the Gaians may appear brutal and anarchic, the really dangerous political conclusion has to do with those who are *outside* of this terrain. According to Latour, "Humans" (as opposed to "Terrans" or "Gaians") are those people who refuse to be "Earthbound," refuse to be political, and presume to be citizens of the world, that is, are "globalists." Because of their rootlessness, humans lack all responsibility to any particular place or soil. They therefore pay no attention to their impacts in the world, or the concrete ecological "feedback loops" that correlate to their actions. In short, while one Gaian people may be a deadly enemy to another, this is always provisional and a matter of political negotiation. But a *human* is worse than the deadliest enemy. They are a borderless, countryless, wandering blight.

> Unlike the Earthbound, Humans are not trustworthy, because you never know where they are headed or what principle marks off the borders of their people [...] Because of this lack of localization, they seem to remain indifferent to the consequences of their actions, postponing the payment of their debts, indifferent to the feedback loops that might make them aware of what they are doing and responsible for what they have done. The Moderns pride themselves on being rational and critical, even while being resolutely non-reflective.[173]

Therefore, the "logical" conclusion of this unbridgeable divide between Gaians and modern humans is a global, civilizational struggle. In Danowski

---

171. Latour, *Down to Earth*, 15.
172. Latour, *Facing Gaia*, 247.
173. Latour, *Facing Gaia*, 251.

and Viveiros's words, there is now the imperative "to make pass [i.e., to abolish] the world of the worldless people."[174] The eco-pessimists deny that such strident localism is the same as the chauvinist, anti-Semitism of Schmitt. Nevertheless, especially in Latour's writings, a direct line is drawn between Mosaic theology and modern decadence.

Latour points to the ancient, Near Eastern practice of "translating" the names of the gods between civilizations. He casts this as a virtuous sort of cosmopolitanism. It allowed for a nontotalizing cultural exchange, wherein different peoples could understand one another's belief systems without subsuming them under any singular theology. In short, translating the names of local gods was an instance of political diplomacy.[175]

This salubrious practice supposedly ended with the advent of Mosaic monotheism. The "Mosaic division" set up *one* God, *one* truth, *one* reality as superior to all other deities. From that point on, local translations were impossible, and everything had to be measured as against this singular Truth. For Latour, modernity has carried on this supremacist legacy. Knowing the one objective Truth about nature has secularized and replaced worship of the one true God. "Whatever we may think of the Moderns, however non-believing they deem themselves to be, however free of any divinity they may imagine themselves, they are indeed the direct heirs of that "Mosaic division.""[176]

Haraway likewise equates monotheisms "in both religious and secular guises," and claims that each tends toward exterminationism.[177] And in Latour's words, both the monotheistic belief in the "Light" and the modern love of "Enlightenment" are synonymous. "Revelation" and "Revolution" share the same destructive belief that one is already living at the end of history.[178] The eschaton has already been immanentized; there are no more secrets or mysteries remaining.

"Humans," "Moderns," or "the people of Nature" (Nature being a globalist term) are therefore incapable of becoming a genuine collective. In possession of their beloved "Truth," any attempt to mark oneself off as just one people

---

174. Danowski and Viveiros de Castro, *The Ends of the World*, 30.
175. Latour, *Facing Gaia*, 155. Latour's critique of modern consciousness as derived from Judaism is indebted to the research of Jan Assmann's books on Moses and monotheism. Assmann's research has proved controversial, with Richard Wolin calling his portrayal of Judaism as a religion of intolerance reductive. See Richard Wolin, "Biblical Blame Shift: Is the Egyptologist Jan Assmann Fueling Anti-Semitism?," *Chronicle of Higher Education*, April 15, 2013, https://www.chronicle.com/article/biblical-blame-shift/.
176. Latour, *Facing Gaia*, 156–57.
177. Haraway, *Staying with the Trouble*, 2.
178. Latour, *Facing Gaia*, 169.

among many can only look like regress. And since modern humans cannot be composed as a specific collective, they likewise are incapable of "occupying the earth" and "defining its territory." For "its universality prohibits it from understanding the relations that it must establish."[179]

However unintentional they may be, certain anti-Semitic tropes here become strikingly obvious. Latour will describe moderns as "ferocious, dangerous, unstable, and [...] profoundly unhappy." More to the point, these people are "wandering souls" given to constant "complaining about the irrationality of the rest of the world." He goes on to describe moderns as being excessively "sensitive" and "in a constant state of anxiety." Above all, their ill-temper is upset by what they perceive as other peoples' irrationalism and relativism.[180] Finally, Latour's prescription for dealing with such uncouth, nervous people is to (above all else) avoid becoming like them: "Even if these people respect no one, we must try to speak to them with respect; this is the only way to struggle against any form of fundamentalism. We must especially avoid imitating their bad manners."[181]

It seems to be more than mere coincidence that modern ills are laid at the feet of Moses, that moderns are described as rootless, rude, neurotic wanderers, hostile to other nations, and that the remedy for modern depoliticization is pinned to the theories of a Nazi jurist. But even if this is all mere coincidence (and again, there is no evidence for *intentional* anti-Semitism), the problem remains.

In his book *Anti-Semite and Jew*, Sartre wrote that "if the Jew did not exist, the anti-Semite would invent him."[182] This claim is particularly relevant here. For the eco-pessimist theories described in this chapter imply a caustic anti-humanism. And if this anti-humanism is couched in a grand political struggle, then there must be at least one people (whatever their specific, ethnic background) which embodies those values that are to be opposed by good Gaians everywhere. This people will necessarily be intellectualist, globalist, and suspicious of local ties and traditions. If they are not also called "the Jews," it hardly matters. They will be painted with the standard reactionary epithets all the same. To paraphrase Sartre: Anti-Semitism is an anti-humanism.

Besides Latour, the other eco-pessimists, including Haraway, Chakrabarty, and Stengers, certainly rely far less upon Schmittian theories of the friend/enemy distinction and the supposed "Mosaic division" of antiquity. All of

---

179. Latour, *Facing Gaia*, 166.
180. Latour, *Facing Gaia*, 166.
181. Latour, *Facing Gaia*, 167.
182. Jean-Paul Sartre and Michael Walzer, *Anti-Semite and Jew: An Exploration of the Etiology of Hate*, trans. George J. Becker (New York: Schocken, 1995), 13.

these theorists, moreover, express some variety of anti-fascist, pro-immigrant, and anti-racist politics. The question then is not one of indicting an individual theorist for their conscious political beliefs. To the contrary, it is one of interrogating whether their basic presuppositions—about Nature, human nature, and social ontology—can hope to sustain a humane politics. But a worldview marked by irreducible differences and agonism (a worldview common to *all* of the eco-pessimists) seems unlikely to achieve that laudable goal.

Chapter 4

COINCIDENCE OF OPPOSITES

In the 2017 film, *Jungle*, Israeli backpacker Yossi Ghinsberg follows a mysterious guide deep into the Bolivian wilderness. The guide launches into an unhinged sermon on the human ego, indigenous societies, and ecological destruction.

> Look at the world. Perfectly balanced. The problem is us. People. We're the cancer. We deserve to disappear, seriously. Communism isn't the answer. Neither is revolution. I tried both. The solution is automation. Mathematical Cosmo politics. Put a freaking computer in charge. No ego trips, no pride. A computer program for the common good.[1]

These words evoke the surprising coincidence of opposites between eco-pessimist and accelerationist thought, between Gaia and Prometheus.

Much of this book has been about highlighting the obvious contrasts between these two intellectual tendencies. The Gaian favors the local, and lives by the maxim that "small is beautiful." They celebrate a sacralized, opaque version of nature which cannot (and should not) be entirely comprehended. Nature (or whatever their preferred term) is beyond our concepts since it is always alive and on the move. Our rightful comportment to nature is not one of exploitation but, instead, deep reverence.

As opposed to this, the Promethean stresses the nullity of nature. Being is less than nothing and can be rearranged to suit the desires of a strong, creative will. Reverence is a holdover of slave morality, and even the "Truth" is no barrier to the hyperstitial imagination.

Nonetheless, these seemingly opposed ideologies (when taken to their logical extreme) find a surprising identity in one another. This culminates in a common anti-humanism, where the human being is always a burden or "drag" on some valorized, superior *other* (whether this be the Earth or A.I.).

---

1. Greg McLean, *Jungle* (Umbrella Entertainment, 2017).

But underneath this shared anti-humanism is a more basic coincidence of metaphysical and ethical positions. It is to these coincidences that we now turn.

Nominalism Is a Noumenalism

Both the Gaian and the Promethean reject the a priori in the traditional sense. They deny that there are any objective, universal standards of knowledge which exist apart from our specific experiences or scientific methods. One way or another, the accelerationist and the eco-pessimist affirm a basically improvisational concept of knowledge formation. Norms of knowing are either generated through our own investigations, or else arise spontaneously from matter itself.

The accelerationist Luciana Parisi made this position clear in her commentary on the mathematical concept known as Chaitin's constant, or "Omega."

> Omega clarifies that randomness is intelligible and detectable within the very computational processing in which unpredictable infinities emerge and operate—and yet cannot be synthesised by an a priori program, theory or set of procedures that are smaller in size than it.[2]

For Parisi, this concept is indicative of a larger ontological point about unpredictability as such. Bruno Latour's Gaian metaphysics echoes a similar commitment with respect to the sciences.

> These sciences must be extended to encompass all processes of genesis, in order to avoid imposing a priori restrictions on the agency of the beings with which we shall have to work. Yet the empirical sciences must also be subjected to certain limits.[3]

Thus, whatever "limits" may exist for empirical research, these will be improvisational and spontaneous too, and not universally imposed. Norms are the offspring of scientific methods or phenomenological experiences, but are *never* considered to be immutable, metaphysical necessities. Even human beings themselves, as the early Haraway argued, are merely a "subsystem" which is localized within an overall architecture that is "probabilistic" and "statistical" rather than necessary.[4]

---

2. Parisi, "Automated Architecture," 414.
3. Latour, *Down to Earth*, 78.
4. Haraway, *A Cyborg Manifesto*, 32.

The immediate consequence of rejecting a priori strictures is to preclude the existence of natural laws. Here, one must be clear in the use of terminology. "Natural law" denotes not merely observed, physical consistencies but rather the idea that the laws of nature logically precede particular objects and events. Obviously, if one rejects all a priori knowledge, then the priority of physical laws over transient events is also abolished.

This is readily seen in accelerationist antinomianism, such as Reza Negarestani's insistence that all "laws" are entirely revisable and are only binding on condition that we commit to them.

> Rationality is the "conception of law" as a portal to the realm of revisable and navigable rules. We only become rational agents once we acknowledge or develop a certain interventive attitude toward norms that renders them binding. We do not embrace the normative status of things outright.[5]

Analogously, Bruno Latour's geo-philosophy rejects "law" as just another illegitimate drag on cosmopolitics. Commenting on his 2015 climate conference (COP21), he states:

> The members of the COP are not *parts* of a higher Whole that would allow them to be unified [...] rather, they are "parties" in the diplomatic sense, in a negotiation that can begin precisely only because there is *no longer a higher arbiter*—neither power, nor law, nor nature.[6]

Here, "law" is purposefully used in an equivocal sense to denote both an international legal regime, as well as universal natural laws. For either sense of "law" would negate the sort of free, improvisational politics that Latour advocates.

If natural laws are denied, then this leads very quickly to a rejection of modern materialism. Since not governed by sovereign, ubiquitous laws, "matter" must, itself, be anomalous. As the *Xenofeminist Manifesto* puts it: "Nothing should be accepted as fixed, permanent, or 'given'—neither material conditions nor social forms."[7] But this accelerationist position is scarcely different from Latour's own notion that our earthbound reality is "changing, local, entangled, and disputable." Both the Gaian and Promethean

---

5. Negarestani, "The Labor of the Inhuman," 455.
6. Latour, *Facing Gaia*, 259.
7. Laboria Cuboniks, *The Xenofeminist Manifesto: A Politics for Alienation*, 1.

stand in contrast to the mechanistic picture of a world as an "external, immutable, universal, and indisputable entity."[8]

Moreover, if matter becomes anomalous, then we can no longer presume that the same rational strictures which govern the human intellect also determine the external world. This lack of isomorphy between mind and nature is, to be sure, an explicit hallmark of much of accelerationist thought. Parochial human experiences don't matter when it comes to defining "the real," which is always inhuman and inorganic.[9]

Yet, even here, a coincidence of opposites presents itself. Exactly by degrading the concept of Nature, and by denying the priority of intelligible laws, "nature" becomes nothing more than the sum total of scientific investigations. We recall Brassier's nominalist scientism:

> Nature is neither more nor less than the various discourses of physics, chemistry, biology, geology, ethology, cosmology [...] The list remains necessarily open-ended. Where the sciences of nature are concerned, the irreconcilable is their highest concept and the irremediable their only meaning.[10]

Thus, a fully nominalized nature becomes surprisingly correlated to the human mind (or, at least, to specific human practices and phenomenological experiences). For science is an inherently human endeavor.[11]

This coincidence of opposites runs in both directions. The eco-pessimist commences their philosophy with a strong sort of correlationism. Nothing at all can exist apart from phenomenological experiences. We may, in some apocalyptic sense, exist apart from the world; but the world never existed before us![12] Nevertheless, it is just this anthropomorphism on the part of the Gaian which (as we have seen in Chapter 3) is incompatible with humanism and intelligibility.

The human essence, since entirely bound up with the whole spontaneous, fluid composition of nature, cannot, itself, be understood. To restate

---

8. Latour, *Facing Gaia*, 287.
9. Brassier, "Liquidate Man Once and for All."
10. Brassier, *Nihil Unbound*, 40.
11. The accelerationist will likely respond that scientific procedures are not based on phenomenal experiences. However, to restate our argument above: If there are no universal, a priori strictures to scientific investigation, then the latter are truly "grounded" in empirical experiences alone. There is no alternative between a priori and a posteriori foundations.
12. Danowski and Viveiros de Castro, *The Ends of the World*, 63.

Viveiros's most telling line, "When everything is human, the human becomes a wholly other thing."[13] And, again, this is hardly different from Negarestani's accelerationist position that humanism "is the initial condition of inhumanism as a force that travels back from the future to alter, if not completely discontinue, the command of its origin."[14]

In the end, noumenalism is a nominalism (and vice versa). It is immaterial where one begins. One may start with a seemingly hard-nosed nominalism which rejects as unimportant our parochial human experiences and any correlation between the human mind and the external world. But since the a priori is rejected, in favor of the strong will, such a nominalism will find nothing stable or intelligible in this external "real" world to hang its hat on. This nominalism will, ultimately, find the external world to be every bit as opaque and mysterious as its noumenalist counterparts.

Conversely, one may begin—as in eco-pessimism—with a noumenal view of reality. In this case, the world is beyond human comprehension. We can only access the world through our active, spontaneous participation in it. "Worlding" replaces knowing. To be sure, for the Gaian, such "worlding" is never the provenance of a singular actor (neither an inventor nor a manager). Nonetheless, we access reality only through our common composition with other creatures. We literally make it what it is, and there are no a priori limits on what we can create together. Therefore, nature is not merely an unapproachable mystery, but rather an opportunity for free-willed creation. It is a species of nominalism after all.

Voluntarism Is an Animism

All of this helps to shape a surprisingly overlapping philosophy of nature. In rejecting the a priori, both the Gaian and the Promethean reject modern notions of materialism. From Spinoza to the French *philosophes* of the eighteenth century, materialism has always been tied to a certain conception of efficient causation. Specifically, there is the idea that the effect is entirely determined by its cause. The world is, in other words, deductively organized—not only in terms of human comprehension, but *really*. Novelty, chance, and spontaneity are but fictions of the human imagination.[15] Or, as Hegel declares in his *Philosophy of History*, "The sole aim of philosophical enquiry is to eliminate the contingent."[16] In its most consistent form, modern materialism allies itself

---

13. Viveiros de Castro, *Cannibal Metaphysics*, 63.
14. Negarestani, "The Labor of the Inhuman," 444.
15. Spinoza, "Ethics," E1P29, E1P33.
16. Georg Wilhelm Friedrich Hegel, *Lectures on the Philosophy of World History*, trans. H. B. Nisbet (Cambridge: Cambridge University Press, 1980), 28.

with Laplace's infamous demon: To have comprehensive knowledge of the present is to predict the future with absolute certainty.

Both the Gaian and the Promethean reject this strong notion of efficient causation, and this is certainly consistent within their underlying antinomianism, that is, their lawless conception of nature. But the question remains: If matter is not governed by a priori laws, then what determines the activity and character of the natural world?

The answer appears to come down, as we have suggested, to the will. For the Promethean, the will is a "destructive," "liberatory," "impersonal" thing which affirms the "non-being of the One," namely, that the world is a chaotic soup.[17] It allows us to shape a nominalist nature as we like. And what's more, the will supposedly transcends any stable conception of human consciousness. In its most extreme variant, not only does Nature become insubstantial, but so does the knowing subject.

Hence, the slogan is "Liquidate man to liberate intelligence."[18] The genius/inventor/CEO cannot be limited by parochial convention, whether social or intellectual. For a certain kind of accelerationist, voluntarism is taken further still: Agency is to be found distributed throughout an increasingly self-aware technology rather than within the human mind at all. For agency cannot be bounded by mere human consciousness.

But this most extreme liquidation (i.e., distribution) of consciousness also marks the Gaian worldview. This derives from a shared intellectual commitment, namely, the "liberation" of effects from their causes. As we saw in Chapter 3, Bruno Latour takes particular issue with modern materialism's tendency to "attribute everything to the causes and nothing to the consequences."[19] In his estimation, it is far more realistic to follow the order of our phenomenological investigations and to give priority to effects (as we experience these first, before any causal story can be postulated). Thus, when Gaians reject modern, efficient causation, they return agency to the effects themselves. The final result is animism, or as Latour puts it, "a prodigious *multiplication* of potential agents."[20]

Like Promethean voluntarism, Gaian animism is dependent on a notion of "free will." Within their metaphysics, it is precisely the notion of "will" which is collapsed into their notion of "action." Having a will is what makes all things potential actors, and vice versa. Latour is clear on this point when speaking of the intentionality inherent to topographical features, such as rivers:

---

17. Brassier, "Liquidate Man Once and for All."
18. Brassier, "Liquidate Man Once and for All."
19. Latour, *Facing Gaia*, 71.
20. Latour, *Facing Gaia*, 163.

Quite to the contrary, there is a will here—that of the competing rivers […] It hardly matters that one [thing] is evoked as having intentionality or will and the other as simply a force, because it is *the tension that makes the actor*, and not the way actors have been endowed with a more or less plausible set of attitudes.[21]

Yet when everything becomes ensouled, active, or alive, this is tantamount to the demotion of individual consciousness. Or, as Morton opines, "consciousness is just another mode of access among equal others."[22] It is not *understanding*, but *doing* that counts and that qualifies one as a genuine agent. And action, taken in its most pristine form, can never be bound to a singular being. The free action precedes the actor, just as the effect precedes the cause. In Latour's words, "waves of action […] respect no borders and, even more importantly, never respect any fixed scale."[23]

But this animistic kaleidoscope is mirrored by the Promethean vision of an unbounded, unchained, creative will. For it is "precisely the type of Promethean project that resists totalization: there is no proper site, nor uniform procedure; it is a generic thought of value creation that formally morphs under localised, material modes of practice."[24] The animism of Latour, Danowski, and Viveiros are mirrored in the cybernetic theories of the early Haraway and Wolfendale, where the locus of agency likewise becomes the "distributed signaling network."[25]

Recent intellectual history pits "strong willed" theorists (Badiou, Zizek, Hallward) as against thinkers of the "distributed will" (Deleuze, Hardt, Negri).[26] This theoretical battleline within contemporary philosophy tends to paint two, highly stereotyped caricatures: The former is of a romantic, adventurous, neo-Leninism, preoccupied with bending history to one's iron will; the latter represents the postmodern dissolution of the will into an ocean of being, multiplicity, and horizontalist anarchism.

Yet as the foregoing discussion should make clear, the parties in this conflict have far more in common with one another than is typically realized. For the will, whether conceived as singular or distributed—voluntarist or animist—is an antinomian thing. It is not so consequential whether free creation is located within rocks, plants, and protozoa, or within inventors, self-driving cars, and

---

21. Latour, *Facing Gaia*, 53.
22. Morton, *Humankind*, 51.
23. Latour, *Facing Gaia*, 101.
24. Reed, "Seven Prescriptions for Accelerationism," 534.
25. Wolfendale, "The Reformatting of Homo Sapiens," 14.
26. Alex Callinicos, *The Resources of Critique* (Cambridge: Polity, 2006), 8.

computer servers. The point is that entities endowed with such a free will cannot truly be understood, nor can they comprehend one another.

This is a problem for both Gaians and Prometheans who want to find revolutionary potential in elements of nature and technology which they interpret as alive, agentic, and free. But the "free will" is not hospitable to political organization or (ironically) to political freedom. There can be no meaningful alliances, and no common projects of emancipation, when there is no common understanding. And there can be no common understanding where there is no common, intelligible law.

Hyperstition Is Humility

Given a worldview marked by the unaccountable will, we must experience reality either through the lens of domination or submission. While the modern rationalist believes that "no one is above the law," the Gaian and Promethean are united in affirming a universe which is always changing, evolving, and unaccountable. Nature must always be something external to or transcendent of the human mind, and it can never be governed by the same rational laws. Since Nature cannot be known as such, it must exist as an alien entity—either to be mastered or obeyed.

The eco-pessimist will insist upon the latter alternative. One must "bow before the majesty of Gaia," rather than attempt to comprehend her (the latter being a form of subjugation and violence).[27] This sounds rather different from the #ACCELERATE Manifesto's declaration that "only a Promethean politics of maximal mastery over society and its environment" can meet today's pressing crises.[28] In its most extreme form, such mastery translates as "hyperstition," or the idea that we can literally form reality through our freely chosen myths.

But in truth, hyperstition and humility are just another coincidence of opposites. The Promethean rebellion against all gods appears to be an act of outright defiance. One insists on obeying no norms except one's own. This rebellion contains a liberatory moment, where all external authority is questioned. Yet that sort of challenge to Olympus is merely negative. It is the assertion of one's own arbitrary will as against the alien will of the other.

For what purpose do we stage such an aristocratic rebellion? This, to the Promethean, is precisely the wrong question. All talk of objective "ends" is already an undue fetter upon one's own spontaneous will. Instead, the

---

27. Latour, *Facing Gaia*, 284.
28. Srnicek and Williams, "#Accelerate: Manifesto," 360.

Promethean rebellion is, "the irremediable form wherein purposeless intelligence supplants all reasonable ends."[29] Nature cannot supply any objective definition of the good for us. It "is indifferent to our existence and oblivious to the 'values' and 'meanings' which we would drape over it."[30] In this sense, Nature is less than nothing, not only ontologically, but normatively speaking as well. *We* are the ones that choose, and we choose for literally no reason.

How, then, do we go about deciding what to make of this insubstantial world? The hyperstitial will is free to create its own mythic narratives. But, again, this freedom is entirely negative. It lacks any conception of real autonomy or agency. It is the "freedom" of the mad king who is bound by no external authority or law, yet can hardly be said to possess a genuine capacity for self-rule.

Freedom, with no a priori limits, turns over into its apparent opposite. The mad king becomes a slave to his own passions. Not only this, but the rule of passions precludes any stable basis for common political accord. Here, one must take Carl Schmitt's geopolitical theories seriously. If the fundamental premise is that the sovereign will freely "decide upon the exception," then the world must truly be a "political pluriverse" with no overarching law or organization. It is a potential war of all against all.[31] Thus, Bruno Latour's denial of any neutral arbiters on the international stage is not incidental to his appreciation for Schmitt's political voluntarism: The two go hand in hand.[32]

But the domestic corollary to this bellicose view of geopolitics is perhaps even more disturbing. Since pristinely unaccountable to any conception of law, the sovereign must neutralize any sense of the political within their own territories. There can be only one sovereign—one political actor—within a given domain.[33] For such a figure cannot be bounded by any common rule or legislation. This point is often missed (or willfully ignored) by liberal readers of Carl Schmitt; but it is part-and-parcel of Schmitt's political voluntarism. The *Führerprinzip*, or leader principle, posits that "the Führer's word is above all written law."[34] As such, constitutional, republican institutions are an anathema to the leader's unbounded will.

---

29. Brassier, *Nihil Unbound*, 47.
30. Brassier, *Nihil Unbound*, xi.
31. Carl Schmitt, *The Concept of the Political*, trans. George Schwab (Chicago, IL: University of Chicago Press, 1996), 53.
32. Latour, *Facing Gaia*, 234.
33. Carl Schmitt, *Political Theology: Four Chapters on the Concept of Sovereignty*, trans. George Schwab (Chicago, IL: University of Chicago Press, 2005), 5.
34. Detlev Vagts, "Carl Schmitt's Ultimate Emergency: The Night of the Long Knives," *Germanic Review* 87, no. 2 (2012): 203.

This dictatorial state would follow, a fortiori, from the premise of a *hyperstitial* will which not only governs the political sphere by decree, but also presumes to govern reality itself through the unhinged power of myth-making. In this tyrannical framework, you are either the dictator, in which case you are a slave to your uncountable passions; or (what is statistically more likely), you are not, yourself, the unaccountable sovereign, in which case you are rendered a powerless supplicant to this alien power. Thus, hyperstition is always a form of humiliation. As Hegel puts it in the *Phenomenology of Spirit*, it is precisely our inability to comprehend the free will which enthralls us to its power. The "unhappy consciousness" is marked by an obedience to that which we cannot know.[35]

As with the other coincidences of Promethean and Gaian thought, this identity of opposites runs in both directions. Not only is hyperstition a humility, but humility turns out to be hyperstitial as well. Gaian humility is centered on the notion that, while Gaia is not strictly comprehensible, our relationship to her is nonetheless value-laden. This is not to be understood in the traditional, moral sense of abstract ethical rules. As Chakrabarty states, "the planetary [...] not only out-scales the human but also [...] has nothing moral or ethical or normative about it."[36] But it is precisely this inconceivable "out-scaling" which makes our orientation to Gaia an affective, qualitative affair, rather than a question of dry facts.

Latour, once again drawing on Schmitt, puts it bluntly: "one cannot make any distinction between facts and values," and so one should reject the modern "bifurcation between natural law and positive law, between *phusis* and *nomos*."[37] Certain accelerationist thinkers would agree. The *Xenofeminist Manifesto*, for example, denies "that the distinction between the ontological and the normative, between fact and value, is simply cut and dried."[38]

However, the most telling theorist on the normative, inhuman qualities of Gaia is, by far, Isabelle Stengers. She describes this primordial figure as "ticklish" and responsive to human actions, yet supremely transcendent and indifferent to human concerns.[39] Threading the needle in this way casts Gaia as

---

35. "Through these moments of surrender, first of its right to decide for itself, then of its property and enjoyment, and finally through the positive moment of practising what it does not understand, it truly and completely deprives itself of the consciousness of inner and outer freedom, of the actuality in which consciousness exists for itself." Georg Wilhelm Friedrich Hegel, *Phenomenology of Spirit*, trans. A.V. Miller (Oxford: Oxford University Press, 1977), 137.
36. Chakrabarty, "The Planet," 28.
37. Latour, *Facing Gaia*, 230.
38. Laboria Cuboniks, *The Xenofeminist Manifesto*, 7.
39. Stengers, *In Catastrophic Times*, 46.

a normative being which, nonetheless, cannot be adequately captured by our merely mortal, ethical categories. Put otherwise, we can't help but care about Gaia, to be her partisan, and to take a political position in her favor; at the same time, we can't ever pin down the objective reasons for our fealty.

It remains an open question *why* one ought to care about this figure which is beyond our comprehension. But the Gaian response is frequently a moralizing one: "How could you *not* care about the Earth and its various critters?" Or as Haraway puts it, "it's not a choice, it's a necessity."[40] Obviously, this is not a satisfactory response, even as it is often compelling on a rhetorical level. What it reveals is a surprising potential for authoritarian thinking, even if not always expressed in political terms. In the absence of ultimate justifications, the only recourse can be an appeal to the passions, that is, "We simply must care."

The Gaians thus seek to compel and cajole their readers as if through sheer force of will. This free use of imagery over reasons ultimately reaches hyperstitial proportions. For even if we cannot truly "know" Gaia, we must still get on with the business of discussing her, writing about her, and forming concrete alliances and political agendas on her behalf. And even if our aim is to "bow before the majesty" of this figure, we require some ideas as to which direction we should lower our gaze. We have to go about inventing Gaia if we are to obey her.

Since the early days of the *Cyborg Manifesto*, a consistent theme in Haraway's writing has been the grounding of politics in myth, rather than in any objective notion of truth. "There is a myth system waiting to become a political language to ground one way of looking at science and technology and challenging the informatics of domination—in order to act potently."[41] The language here is starkly voluntaristic, where we freely compose our own realities in order to mount a political program. This idea carries through to her more recent book titled *Staying with the Trouble* where Haraway speaks of the need to "become more ontologically inventive" and uses the initials "SF" to elide the notions of "science fact," "science fiction," "speculative feminism," and "speculative fabulation."[42] Likewise, Stengers will suggest that all scientific interpretation requires "artifice," "fabrications," and "invention," far more than "the truth."[43]

We therefore find that Gaian humility and Promethean hyperstition are not so opposed as appearances suggest. To the contrary, a humility based in the lack of objective knowledge will soon overturn into a hyperstitial arrogance.

---

40. Haraway, *Staying with the Trouble*, 73.
41. Haraway, *A Cyborg Manifesto*, 66.
42. Haraway, *Staying with the Trouble*, 98, 2.
43. Stengers, *In Catastrophic Times*, 146.

Such arrogance believes that it can freely create the object of its worship, and freely demand that others submit to its creation. This, as we have seen, mirrors the Promethean paradox, whereby a hyperstitial will becomes enslaved by its own artifice and passions.

Mystery, Magic, and Faith

These coincidences of opposites are only possible because they each participate in a more abstract category. *Mystery*, *Magic*, and *Faith* are the common ideas which unify Gaian and Promethean thought. If nominalism is a noumenalism (and vice versa), then this is because they are both instances of mystical thinking. That is to say, they both posit some object which is opaque to the human mind, and thus unknowable. For noumenalism, even of the "blackbox" variety, unknowable entities are the everyday objects that we encounter. These are irreducibly bound up in problematic social relations and phenomenologies which investigation can never fully untangle. Nominalism *appears* to do away with the opaque, rich lifeworld of the Gaian.[44] But in negating the substantiveness of Nature, that is, in making it less than nothing, it affirms that reality is wholly unknowable. The categories of the human mind do not apply to the outside world. Despite their atheistic bluster, Promethean nominalism turns out to be a negative theology after all.

Voluntarism is an animism (and vice versa) because each is a form of magical thinking. Since reality cannot be adequately known, our relationship to the world must be one of unregulated action. Natural laws do not apply to a mystical universe, at least not in a ubiquitous, i.e., genuinely "lawful," way. Instead, "world" becomes a verb, à la Haraway's concept of "worlding." We *world* our environment, and not according to any preset plan or essence. For the Gaian, this is always cast as a collaborative, animist pursuit, while the Promethean often promotes the voluntarist image of the tortured, lonely creator. In both cases, action in the world is magical because it is free in a negative sense, never bound by objective limits or reasons. For all their invocations of materialism, the eco-pessimist and accelerationist are not materialists, but magicians.

Finally, hyperstition is humility (and vice versa) because each involves a profession of faith. Certainly, "faith" is not a word that is typically associated with Promethean bravado. Their program is one of desacralizing and instrumentalizing everything. The free creator engages in hyperstition—a

---

44. Ruether, *Gaia & God*, 237–38.

rhapsodic creation-for-its-own-sake, neither tied to natural laws, "normal" science, or the objective ends of human welfare. In doing so, the creative will is so liberated as not to be bound, even to the creator themselves. Thus, in many accelerationist texts, artificial intelligence, technology itself, or even free-floating capital, are the disembodied loci of the inventive will. The will finally becomes transcendent, alien—and perhaps hostile—to the human frame. But this is shockingly similar to a Gaia which is freely composed or "worlded" by the assemblage of living critters. She too is spoken of as a transcendent, indifferent figure to whom we must bow down. Faith, if it has any stable meaning at all, means precisely this sense of fealty beyond all reason.

This coincidence of opposites is readily cashed out in mythological terms. Gaia, as Hesiod describes in the *Theogony*, is the "ever-firm foundation of all."[45] This is an appropriate description given Gaia's identification with the Earth. Nonetheless, even for Hesiod, Gaia is an anomalous figure with entirely obscure origins; at one moment there is "only Chaos, the Abyss," and the next moment Gaia just appears for no particular reason or cause.[46]

Gaia's origin is akin to the "uncaused" cause that is the theistic idea of God, as opposed to the self-causing (*causa sui*) Substance of Enlightenment philosophy.[47] Both are, in their own ways, independent beings. But crucially, the former remains a mystery since its reason for being is always obscure and anomalous. Something ineffable always remains in reserve. Enlightenment Substance, since self-caused, is not only rigorously governed by its own laws, but this self-legislation is also the template for the ordinary laws of nature. In Hesiod's account, Gaia is clearly the former sort of being—entirely spontaneous in her origins and unable to sustain a consistent natural law.

Like Gaia, Prometheus is described in antinomian terms. In Hesiod's tale, Prometheus functions as the arch-rebel, undermining the Olympian rule of Zeus. In stealing celestial fire, and in cheating the sacrificial rites, Prometheus thumbs his nose at order itself. To be sure, Gaia's opposition to the law comes from a different perspective; she is *anterior* to the rule of Zeus and the Gods. Prometheus (though, himself, a titan) always plays the part of the trickster upstart, and so appears to be a youthful challenger. But what they have in common is that they threaten to plunge established, hierarchical order back into chaos.

---

45. Hesiod, *Works & Days. Theogony,*, Theogony 116–18.
46. Hesiod, *Works & Days. Theogony*, Theogony 116–18.
47. Spinoza, "Ethics," E1D1.

The Olympian Reaction

The idea of a crumbling world order is every bit as relevant to today's late capitalism, as it was to Hesiod's Iron Age. A certain strain of contemporary conservatism has become particularly obsessed with this perceived threat. It is born of anxieties about civilizational decline, mindless consumerism, secularism, the breakdown of the family, and social levelling. This idea of the *Chaoskampf* (the struggle against chaos) is a continuation of themes from both ancient and medieval lore.[48]

Today, the anxiety about an ever-encroaching chaos takes the form of pop-psychology and self-help sermons, often directed at young men.[49] Jordan Peterson's dictum that one must "set one's house in perfect order" before trying to change the world fits this pattern.[50] So too is Navy Admiral William McRaven's insistence that, "If you want to change the world, start off by making your bed."[51] This obsession with personal and familial order would be well-appreciated by Hesiod who, you will recall, ties the hierarchical rule of Zeus to the equally stratified order of the Greek household and society.

Peterson cites two significant threats to order: The first is that of the all-devouring mother. Here, it's hard not to think of Gaia, before the rule of Zeus. But more fitting still is the Babylonian goddess Tiamat, a primordial, oceanic symbol of chaos who has to be dismembered by the hero Marduk in order for civilization to flourish.[52] For Peterson, today's "social justice warriors" and "postmodern feminists" incarnate this devouring mother archetype. They treat all people as either wicked predators or as helpless infants, thereby

---

48. For the origin of the Chaoskampf concept, see Hermann Gunkel's classic 1921 study, *Creation and Chaos in the Primeval Era and the Eschaton* (Grand Rapids, MI: William B. Eerdmans, 2006).
49. Karen Heller, "Jordan Peterson Is on a Crusade to Toughen up Young Men. It's Landed Him on Our Cultural Divide," *Washington Post*, May 2, 2018, sec. Style, https://www.washingtonpost.com/lifestyle/style/jordan-peterson-is-on-a-crusade-to-toughen-up-young-men-its-landed-him-on-our-cultural-divide/2018/05/02/c5bafe48-31d6-11e8-94fa-32d48460b955_story.html.
50. Jordan B. Peterson, *12 Rules for Life: An Antidote to Chaos* (Toronto: Random House Canada, 2018), 147.
51. Navy Admiral William H. McRaven, "Address to the University of Texas Commencement" (University of Texas at Austin, May 17, 2014), https://www.youtube.com/watch?v=yaQZFhrW0fU&t=1s.
52. Jordan B. Peterson, *Maps of Meaning: The Architecture of Belief* (London: Routledge, 1999), 123.

criticizing every sign of strength and virility, while keeping the disadvantaged in a prolonged state of dependency.[53]

But the other threat to order, in this conservative imagination, is that of the defiant figure of Satan—a prideful trickster akin to Prometheus. So the threat is not only one of motherly suffocation, but also the reckless subversion of a cynical know-it-all. Satan is the consummate antihero; where the "world-creating exploratory hero" is humble before the mysteries of an unknown world, the Devil is marked by a totalitarian pride in his own knowledge, pretending to be capable of God's own omniscience. The tragic figure of Ahab, in Melville's *Moby Dick*, is therefore described as both demonic and Promethean, driven to pierce the white whale, the symbol of unattainable knowledge.[54]

Neither Gaian surrender nor Promethean hubris are accepted by this sort of reactionary pundit. The proper disposition to mortal life is to recognize our inherent limitations, as well as the enduring imperfection of the world at large. We should confront the unchangeable fact that "pain and anxiety are an integral part of human existence." At the same time, we are to go on struggling against this existential entropy in the hopes of carving out local islands of order and satisfaction, that is, to discover a "system of value" which can stave off nihilism, hopelessness, and despair.[55]

The upshot of such conservatism, then, is a stubborn dualism which rejects all final resolutions. The lawless void will always threaten concrete order, but that is no excuse for giving up. In this Manichean scheme, we are always compelled to fight against the devil. Neither, however, should we presume to thoroughly tame reality, and so impose a suffocating, equalitarian regime or nanny state. For such utopianism is doomed to fail and will only reproduce the disorder and chaos which it seeks to expel. For Peterson, the "postmodern feminist" and "SJW" are corrosive to all established order; their efforts at levelling can only result in a relativistic hell.

This conservative stance is anticipated in Mary Shelley's *Frankenstein* which, relatedly, ends with Victor's lament of ever having embarked on his quest for knowledge of life and death.[56] The gothic horror of H. P. Lovecraft strikes a similar tone. In *The Dunwich Horror* (1929), he warns against the forbidden knowledge of the *Necronomicon* which summons uncontrollable alien powers.

---

53. Jordan B. Peterson, "Strengthen the Individual: Q & A Parts I & II" (Ottawa Public Library, Ontario, Canada, March 11, 2017), https://www.youtube.com/watch?v=_UL-SdOhwek&t=0s.
54. Peterson, *Maps of Meaning*, 296.
55. Peterson, *12 Rules for Life*, xxxi; Steven Pinker, *Enlightenment Now: The Case for Reason, Science, Humanism, and Progress* (New York: Viking, 2018), 23.
56. Shelley, *Frankenstein or The Modern Prometheus*, 2003, 220.

Once again, trying to tame the unknowable void always backfires. As Professor Armitage, the hero of the story, admonishes, "We have no business calling in such things from outside, and only very wicked people and very wicked cults ever try to."[57] Lovecraft's position is stated all the more explicitly in his most famous story *The Call of Cthulhu* (1928).

> The most merciful thing in the world, I think, is the inability of the human mind to correlate all its contents. We live on a placid island of ignorance in the midst of black seas of infinity, and it was not meant that we should voyage far. The sciences, each straining in its own direction, have hitherto harmed us little; but some day the piecing together of dissociated knowledge will open up such terrifying vistas of reality, and of our frightful position therein, that we shall either go mad from the revelation or flee from the deadly light into the peace and safety of a new dark age.[58]

The Olympian wants the secure, provincial island of limited knowledge surrounded by a sea of mystery, and no more. In mythological terms, they want Zeus and Olympus, and the whole social order that they prefigure, but not the total demystification of Nature.

However, this conservative position does not escape the gravitational pull of that which it rejects. It sets up Olympian order as against two forms of disorder and lawlessness—the Promethean and the Gaian. But what is the rule of Zeus if not an unaccountable tyranny? Zeus is, himself, above the law. In defeating Gaia and punishing Prometheus, he claims to bring security, stability, and logos to the world. And yet, there can be no real security—and no actual Reason—if the sovereign is always he who decides upon the exception. With such an unaccountable authority, the sovereign is just another player in the war of all against all. If Olympian order is the template for earthly politics, then we are left in a state of nature with merely a kingly veneer.

---

57. H. P. Lovecraft, "The Dunwich Horror," in *H. P. Lovecraft, The Complete Fiction* (New York: Barnes and Noble, 2008), 667.
58. H. P. Lovecraft, "The Call of Cthulhu," in *H. P. Lovecraft, The Complete Fiction* (New York: Barnes and Noble, 2008), 355.

Epilogue

BEYOND THE VOID

Can there be a real alternative to this antinomian picture? One which avoids the chaos of an unknowable Nature, or a lawless rebellion, or an unaccountable sovereign? Perhaps there can be such an alternative, but only if we get to the metaphysical root of the problem itself. What the Gaian, Promethean, and Olympian positions each share is a fundamental dualism between reason and its other, what we may call "the void." Each formulation posits an outside to intelligibility. There is something we can never know. And in this void, we find monsters. Thus, we are left with the choice of embracing the monstrous in one way or another (Prometheus and/or Gaia), or else reaching an always provisional ceasefire with the encroaching chaos (Olympus).

But what if there were no outside to reason? What if, instead, Nature is a thoroughly intelligible substance? In that case, the laws which govern the intellect would also mirror the order-and-connections of material reality. This is certainly an unpopular position in contemporary philosophy, thoroughly infused as it is by postmodern, existentialist, and pragmatic assumptions. Still, truth is no popularity contest. And this rationalism has much to recommend it.

First, there is the oft-repeated Hegelian insight that "to posit a finite limit is to go beyond that limit."[1] All boundaries to knowledge are ultimately provisional and relative. In other words, even our local instances of ignorance are conditioned by the contours of a more broadly intelligible world. We have to ask *why* and *under what conditions* we fail to comprehend something. Otherwise, it wouldn't be possible, in the first place, to identify our specific deficits of knowledge.

Lovecraftian monsters are the supreme rejection of this Hegelian insight. For Lovecraft, these beings are threatening precisely because they cannot be known. They strike the senses without being understood.[2] But for the

---

1. Georg Wilhelm Friedrich Hegel, *Science of Logic*, trans. A.V. Miller (New York: Humanities Press, 1976), 93. 133.
2. For epistemological considerations in Lovecraft's work, see Graham Harman, *Weird Realism: Lovecraft and Philosophy* (Winchester, UK: Zero Books, 2012).

rationalist, we have a simple disjunctive syllogism at work: Either we do not comprehend the monster at all, in which case it is precisely of no consequence to us, or else, we do understand something of the beast, in which case it is merely a question of piecing together additional insights. Whichever is the case, the *mysterium tremendum* (and the humility of mystical thinking that goes with it) is overcome. In the rationalist's thoroughly demystified world, there are only cryptids which have yet to be classified; there are no monsters.

At the base of this rationalism is the originary claim that we do possess some knowledge. This is not, to be clear, the outsized assertion that we know every finite thing all at once (a common, but misguided, cartoon of the rationalist). Instead, it is merely the position that our sundry, empirical observations are grounded in at least one, certainly true idea—a bedrock of knowledge. This is the opposite of a skepticism which categorically rejects any certain knowledge whatsoever.

Whereas academic skepticism seems to undercut its own claim ("I know that I know nothing" is a self-contradiction), a more sophisticated variant of skepticism comes from the Pyrrhonic tradition. Here, the claim is instead that one is never certain of anything, including whether or not we have any knowledge. While much ink has been spilled in praise (or fear) of this allegedly more durable skepticism, Spinoza himself seems to have been suitably unimpressed.

In his *Treatise on the Emendation of the Intellect* (1677), Spinoza points out that an unwillingness to claim certainty is little more than a paralytic. The skeptic, by their own standards, ought to remain in perfect silence:

> If they affirm or doubt something, they do not know that they affirm or doubt. They say they know nothing, and that they do not even know that they know nothing. And even this they do not say absolutely. For they are afraid to confess that they exist, so long as they know nothing. In the end, they must be speechless, lest by chance they assume something that might smell of truth [...] If they deny, grant, or oppose, they do not know that they deny, grant, or oppose. So they must be regarded as automata, completely lacking a mind.[3]

More to the point, a Pyrrhonic skepticism is every bit as self-refuting as its more basic, academic counterpart. For knowledge is simply defined as "having a certainly true idea." Thus, to declare that one is always unsure of whether one possesses knowledge (again, a certainly true idea) is to refute oneself. This

---

3. Benedictus de Spinoza, "Treatise on the Emendation of the Intellect," in *The Collected Works of Spinoza, Volume I*, ed. and trans. Edwin Curley (Princeton, NJ: Princeton University Press, 1985), 22.

is tantamount to saying that "I am always unsure about whether I am sure about anything."[4]

This, in the end, is why any limit to knowledge must always be provisional. We do, necessarily, possess some certainly true ideas (supposing we have any ideas at all). As such, all confusion and ignorance must be understood in this context: not one of total confusion, but of a background of certainty.

Nature versus Mystery

But what work does this "certainly true idea" actually do for us? In the rationalist tradition, sure knowledge is the basis for doing "first philosophy," or in other words, for deducing our fundamental knowledge of *what is*. This is different from later, critical philosophies which seek to foreground "method" before any substantive knowledge claims. For Spinoza, Hegel, and other rationalists, however, a purely formal method cannot be primary. This would lead to an infinite regress whereby a formalistic law or rule has to be guaranteed by some other second-order rule, and so on ad infinitum.

> The first thing we must consider is that there is no infinite regress here. That is, to find the best Method of seeking the truth, there is no need of another Method to seek the Method of seeking the truth, or of a third Method to seek the second, and so on, to infinity. For in that way we would never arrive at knowledge of the truth, or indeed at any knowledge.[5]

What is required, instead, is an indubitable idea which can ground method, namely, our formal laws and procedures.

> From this it may be inferred that Method is nothing but a reflexive knowledge, or an idea of an idea; and because there is no idea of an idea, unless there is first an idea, there will be no Method unless there is first an idea. So that Method will be good which shows how the mind is to be directed according to the standard of a given true idea.[6]

---

4. This claim is uniquely self-refuting in a way that other claims of uncertainty are not. For example, to claim that "I am always unsure of whether there is a cake in the refrigerator" is not self-refuting. We may, indeed, always lack such knowledge just in case we never open the fridge door. But to say that we are always uncertain of having any certainty is, clearly, different. The content of this proposition's terms renders it uniquely self-contradictory and incoherent.
5. Spinoza, "Treatise on the Emendation of the Intellect," 16.
6. Spinoza, "Treatise on the Emendation of the Intellect," 19.

Hegel makes an analogous point in describing method as "the Notion [Concept] that is determined in and for itself." And this Concept is described by Hegel as both "concrete" and "self-knowing."[7]

If we do have a certainly true idea, then it stands to reason that this thing must be known through itself. Otherwise we lapse into the same problem of an infinite regress: We know "A" through "B," and "B" through "C," and so on, without end. If, on the other hand, we know a thing through *itself*, then this hypothetical object must also exist independently. Hence, Spinoza's definition of "substance" in part 1 of his *Ethics*: "what is in itself and conceived through itself."[8]

That is why, in Spinoza's *Ethics*, simple "substance" is necessarily self-causing (*causa sui*). Since it is independent, no other thing can create or condition it. And in causing itself, without any external limits, it expands to become the whole of infinite "God or Nature." To be clear, this is a naturalistic pantheism, and not a theism. "God" denotes here only an infinite, material, intelligible substance.[9] Once this starting point to philosophy is established, we have the kernel of a subsequent "philosophy of nature," an approach to the sciences that is entirely free of miracles, superstitious free wills, and fundamental indeterminacy of all sorts. Since our originary object was intelligible, then intelligibility itself expands to fill the whole of existence. Mystery is pushed out entirely.

---

7. Hegel, *Science of Logic*, 827. Some may object that we are eliding the differences between Spinoza and Hegel; that Spinoza is typically cast as an abstract rationalist, without a developed philosophy of history, and Hegel is the philosopher of history par excellence. But this obscures the fact that Spinoza does not ignore or discount history in the *Ethics*, or the fact that Hegel himself is a metaphysical rationalist. It is true that Hegel is responsible for the mischaracterization of Spinoza as an ahistorical Parmenidean, but the authors have argued why this is a faulty interpretation in various works. See, for instance, Harrison Fluss and Landon Frim, "The Wrong Couple," *Radical Philosophy*, June 2018, https://www.radicalphilosophy.com/reviews/individual-reviews/the-wrong-couple; Harrison Fluss, "The Specter of Spinoza: On the Legacy of the Pantheism Controversy in Hegel's Thought" (Doctoral Dissertation, Stony Brook, NY, Stony Brook University, 2016).
8. Spinoza, "Ethics," E1D3.
9. Despite Spinoza's theological-sounding terminology, his rationalism is constitutionally opposed to all theism because the latter presumes an entirely free, often occult and numinous, being. But the import of Spinoza's self-causing substance is precisely intelligibility and determinism. Thus, while Spinoza's God may be said to have "personality" (in the Hegelian sense of being self-moving and self-differentiating), it is decidedly not a "person." See Spinoza, "Ethics," E1P11; E1P14.

Law versus Magic

Unlike the essentially magical worldviews of both the Promethean and Gaian, our philosophy of nature is a proper materialism. The crucial distinction here has to do with the priority of substance over its various effects. We recall that, for Prometheans and Gaians alike, there is the tendency to reject this priority, and so to free the effects from any overarching first cause. For instance, there is Latour's reversal of cause and effect or Negarestani's denial of "natural necessity."[10] In rejecting causal determinism, Prometheans and Gaians both emphasize unbounded, spontaneous, creative action, freed from all intelligible necessity. They paint necessity and substantial first causes as merely theological holdovers.[11]

But a genuine materialism realizes that the ontological priority of substance is the furthest thing from theism. For if substance is prior to its effects, then natural laws are truly unbreakable. Not only does "God not play dice," but God is also deprived of "His" traditional, theistic free will. There are no miracles or fissures in the natural order of things and thus no opportunities for *creatio ex nihilo* or divine intervention. For this would presume God changing himself, or at least changing his mind. Similarly, the laws of substance must remain constant. Particular effects, modes, and events do change, but the laws of motion do not; they are part of the eternal, self-causation of Nature as-such.[12] In other words, the laws of motion determine how things move, but are not themselves moved.

---

10. Latour, *Facing Gaia*, 68, 69; Negarestani, "The Labor of the Inhuman," 452.
11. This tendency is manifest in Peter Wolfendale's emphasis of neo-Kantian, normative thinking over what he characterizes as an "Aristotelian," that is, metaphysical, preoccupation with material causes. See Wolfendale, "The Reformatting of Homo Sapiens."
12. This is basic to Spinoza's distinction between *Natura naturans* (or patterning Nature) and *Natura naturata* (or patterned Nature). Natural laws are distinct, and have priority over, the instances of physical reality that they condition. At this point, some may wish to invoke certain views of quantum mechanics (particularly the so-called Copenhagen Interpretation) so as to confute the above claims of determinism in Nature. While we certainly do not wish to limit empirical investigation in any way, it should be noted that the very same set of empirical data has resulted in a plethora of interpretations among theoretical physicists and philosophers of science, not all of whom accept radical indeterminacy as an "objective" feature of reality. More to the point, however, is a basic question of method: Are empirical observations taken to be prior to all conceptual, a priori, categories and strictures? If we go that route, then we are left with the absurdity that we can think or express to one another the results of empirical experiments wholly independent of concepts and logic. What Spinoza is up to in the *Ethics* is the creation of a "first philosophy" that uses conceptual analysis in order to derive apodictic conclusions. Whether this philosophy succeeds or fails is not a question of measuring it against empirical data. Moreover, Spinoza's deterministic conclusions should not be misinterpreted. He does not rule out "conditionality" in the common sense of the

Freedom versus Faith

This determinism may appear to be entirely bloodless and amoral. What could be less politically motivating than a "mechanistic" universe of orderly actions and reactions? But this is to confuse a substantive rationalism with its pale, positivist imitation. For the genuine rationalist, finite causes and effects are not the whole story. These are, we have seen, always conditioned by a singular Substance which is prior to them all. And so it is through this one Substance that all finite creatures derive their being and identity.

As against all species of atomism, the rationalist asserts that I cannot begin to adequately comprehend my own finite nature without reference to the infinite Nature of which I am a part. The corollary to this is that my self-conception is always tied to the thought of other beings. For they, likewise, share in the same substantial reality as me. At its apotheosis, this recognition of identity with other creatures becomes an expanded sort of self-love. This idea is far from new and is well-expressed in the ancient Stoic concept of "oikeiôsis," that is, the expansion of oneself to include others. In modern philosophy, it is manifest in Spinoza's correlation of personal striving (*conatus*) with the intellectual love of all.[13]

But what confounds most readers about this rationalist ethics is not its universalism, so much as its utter necessity. There is, and can be, no free moral *choices*. We always do what we think is right. This seems perplexing for two distinct reasons. First, we have been conditioned to think of ethics as a series of decisions, and in the absence of free choice, one typically feels that there can be no morality. It is for this reason that we don't think of computer programs or houseplants as moral agents. Second, we rightly perceive the countless examples of irrational, hateful, and unethical behavior which mar contemporary society. Nothing about ethics, in other words, seems "necessary" at all.

But the rationalist does not presume that human beings are angels, sublimely free from the passions. One can reject original sin but still recognize the pull of ordinary, worldly appetites.[14] What distinguishes the rationalist is simply the recognition that insofar as we do recognize what is right, and have the power to act on this information, then we will necessarily do so. This is

---

word, as though each thing depends on God-as-such individually. Rather, natural laws act as the "substrate" which condition the transient interactions between finite things. Event B depends on Event A, and both are determined (*sub specie aeternitatis*) by the whole set of physical laws.

13. Spinoza, "Ethics," E4; Landon Frim, "Impartiality or Oikeiôsis? Two Models of Universal Benevolence," *Symposion* 6, no. 2 (2019): 147–69.
14. Indeed, an entire section of Spinoza's magnum opus is dedicated to a meditation on the passions as a source of human bondage. See Spinoza, "Ethics," E4.

the Platonic rejection of *akrasia*, namely, the "weak will."[15] One may falter, but never because of some uncaused desire to be wicked for its own sake. We may be ignorant of the true import of our actions (as with the imperceptible buildup of arterial cholesterol from fast food), or we may only *appear* to be acting when in fact our body is being mechanically compelled (as with an overpowering, chemical addiction). In either case, we never knowingly choose what is bad.

Even in the less obvious case of speeding through a school zone, it is not as if the driver is acting for no reason at all, or even more perversely, through a spontaneous evil will to do harm. Rather, the speeder momentarily "backgrounds" the possibly tragic implications of their poor driving while their appointment across town is front of mind. We always aim at some perceived "good," even when we act badly.[16]

Combine this with the insight of substance monism, and we have the foundation for a universal politics of solidarity and emancipation. In recognizing our identity-in-Nature with others, their welfare is necessarily counted as one with our own. We do not treat our neighbor *as if* they were us; in some real sense, for the consistent rationalist, *they are us*. Just as we necessarily seek our own advantage and well-being, so we seek the welfare and flourishing of others with the same necessity.

Monism, in securing an intelligible reality, therefore accomplishes two things at once: It allows for the instrumentalization of nature in order to improve the lot of sentient and sapient beings. For only an intelligible and unified nature can be purposefully modified to meet particular, desired ends

---

15. G. M. A. Grube, *Plato: Five Dialogues: Euthyphro, Apology, Crito, Meno, Phaedo* (Indianapolis, IN: Hackett, 2002), 30.
16. One might object with the counterfactual example of an individual who speeds through a school zone with the clear idea (or even hope) that they strike an innocent child. Comic book super-villains come to mind here, though we do not doubt that real-life instantiations also exist. But in such extreme cases it is even clearer that we are not witnessing sober, deliberate, "free" decisions, but rather behaviors which are, in fact, determined by severe emotional and mental imbalances. Merely claiming that one "clearly and rationally" desires to cause suffering to others is not sufficient evidence that this is true. Substance monism informs us that the killer is of the same substantial identity as his victim, and so (if rational) must care for their welfare and suffering as they do their own. That this compassion does not, in fact, always occur is no indictment against first philosophy, but merely evidence of a breakdown in the criminal's ability to reason clearly. The mere subjective claim that 2+2=5 does not undermine the fundaments of arithmetic, and similar professions in the moral sphere (which one can readily find in the words of a Sade, Baudelaire, or Dostoevskian antihero) should be given no greater weight as evidence.

(think crop irrigation, synthesizing medicines, or building shelter). But at the same time, monism provides the reason why we necessarily strive to do this work in the first place. Insofar as we are rational, we *will* pursue our welfare in solidarity with others who share in our same nature.

This is the outright rejection of both Gaian and Promethean conceptions of faith. For eco-pessimists and accelerationists each presume some superiority of values over facts. And given this independence of the normative, values become a matter of faithful affirmation rather than objective knowing. In the case of the Gaians, this primacy of value is correlated to an occulted nature which cannot be thoroughly understood, but nonetheless ought to be obeyed faithfully. For all their talk of "nature-culture," the "facts" of nature don't really exist in any objective way. With the Prometheans, it is a stark, valueless world where all normativity is a free, hyperstitial creation. Values are supreme because *we* impose them.

But these professions of faith (whether directed at Gaia or our own will) are incapable of sustaining an emancipatory politics. For nothing stable grounds our caring about the other, much less our substantial identification with them. Certainly, given their common denial of a universal human nature or essence, nothing grounds our political solidarity with those from entirely distinct cultures or geographies. For who is to say that *their* freely chosen values, namely, *their faith*, will be compatible with our own? Finally, the only path out of this subjectivist contest of wills is the return to an objective metaphysics; and not only this, but a *monistic* metaphysics in particular—one which insists on the substantial unity of all peoples.

Myth Moving Forward

This book has been an analysis of two mythic narratives, the Gaian and the Promethean. It is through their respective lenses that so many people today understand our most pressing crises. Do we solve the challenge of climate change by embracing the mantra of "small is beautiful"? Or must we cede control to world-transforming technologies? Will humility or hyperstition save us?

In the end, Prometheus and Gaia represent a strange coincidence of opposites. They each embrace the inhuman, and seek salvation through some obscure, transcendent object (whether nature or technology). But we ought not to think that this escape into the transcendent is a necessary consequence of myth itself. Human beings have a natural tendency to illustrate our most potent ideas through poetic imagery. That is the essence of myth-making, and we should not relinquish this vital, human practice to the most mystical among us.

The only difference between healthy myth-making and mysticism is the degree to which one is clear-eyed about our instrumental use of imagery. The image must be understood as posterior to the Concept; it is a useful tool, and nothing more. Once we grasp this fundamental point, there is no need for anxiety when it comes to using the figures of Prometheus and Gaia (or any other mythic figure for that matter). For they needn't necessarily represent the antinomian, voluntarist worldviews of the contemporary accelerationist or eco-pessimist. *We* choose what these images can signify.

Moreover, there are literary precedents for a more enlightened use of these titanic symbols. While Hesiod was preoccupied by Late Bronze Age anxieties over loss of control and order, the later Aeschylus took a brighter view. Writing during the Golden Age of cosmopolitan Athens, Aeschylus saw far greater capacity for human achievement. Thus, in Aeschylus's *Prometheus Bound*, the titular character opposes a tyrannical Zeus, not only for his own egotistical sake, but more clearly for the benefit of humanity itself. In addition, this version of Prometheus is not starkly opposed to the Earth or Gaia, but calls out to her for support.[17]

Such a humanistic reading was given modern expression in Percy Shelley's *Prometheus Unbound* (1820). Here, Shelley repeats Aeschylus's depiction of Prometheus as a savior on behalf of humanity. Two key differences obtain, however; each illustrates Shelley's moral and political advance over Aeschylus. The first is that there is no final reconciliation (as with Aeschylus) between the rebel Prometheus and the tyrannical Zeus (Jupiter). All unaccountable Olympian authority is abolished. Second, and perhaps even more important, is that Prometheus's antiauthoritarian mission to overthrow Olympus is aided by a novel figure Shelley calls the "Demogorgon." This "people monster" has been widely interpreted to represent the masses, especially when imbued with an energetic, revolutionary spirit.[18] Humanity finally takes center stage.

The closing lines of *Prometheus Unbound* involve a surprising speech delivered by the Demogorgon to Prometheus himself. It is surprising because, while laudatory, it almost takes the form of a lecture. This humble "people monster," which began as an ignorant, cave-dwelling beast, now instructs the great Titanic savior on the true virtues of triumphant, post-Olympian existence: "To suffer woes which Hope thinks infinite; To forgive wrongs darker

---

17. Philip Vellacott, *Aeschylus: Prometheus Bound and Other Plays* (London: Penguin Books, 1961), 280–301; references are to the line number.
18. Donald H. Reiman and Neil Fraistat, eds., *Shelley's Poetry and Prose* (New York: W. W. Norton, 2002), 206–7. On the echoes of the American and French Revolutions in Shelley's epic poem, see Gerald McNiece, *Shelley and the Revolutionary Idea* (Cambridge, MA: Harvard University Press, 1969), 242.

than death or night" is the mark of the enlightened being. These virtues are not limited to the personal, but take on a decidedly political tone as well. "To defy Power, which seems omnipotent [...] Good, great and joyous, beautiful and free; This is alone Life, Joy, Empire, and Victory."[19]

We may take our cue from this more progressive lineage of mythological writing. It is up to us to employ these symbols so as to illustrate a truly rationalist worldview and so to inspire a genuinely emancipatory politics. This will involve emphasizing the reconciliation of Prometheus and Gaia, and crucially, insisting that this reconciliation is achievable through the figure of the human.

We can affirm Marx's vision of Prometheus as the enemy of illegitimate authority everywhere, a rebel "against all heavenly and earthly gods" who recognizes only humanity "as the highest divinity."[20] Prometheus evokes the liberation of technology and production from the rock that is class society.[21] It is the abolition of an artificial scarcity that is imposed, not by natural circumstances, but by the profit-motive. The masses must become like Prometheus unbound, just as Prometheanism must become social and democratic. It is a matter of taking the fire of technology into our own, collective control.

At the same time, this collective control is nothing, and remains utterly directionless, if we imagine humanity as merely a free-floating subjectivity. If we persist in the delusion that there is no human nature, no determinate human needs, or conditions for flourishing, then all the agency and control in the world will have no meaningful purpose. It is the fact of our common identity *in Nature*, indeed that we are all "a part of nature," that makes us one humanity sharing a common essence.[22]

In this, Gaia should not be understood as a transcendent, mysterious being, but rather an intelligible Substance to which we all belong. This objective identity not only reveals the intelligible conditions for our flourishing, but as we have also seen, actively compels us to pursue this common good.

But this cannot be realized without a transformation of the whole social order. The overturning of class society is necessary to finally reconcile the

---

19. Reiman and Fraistat, *Shelley's Poetry and Prose*, 285, 286. There are residues of hyperstitial thinking even in Shelley's romantic prose. The point, however, is that it is up to us how we instrumentalize such received imagery.
20. Dirk J. Struik and Sally R. Struik, trans., *Karl Marx and Frederick Engels, Collected Works*, vol. 1 (New York: International, n.d.), 30.
21. Karl Marx, *Capital: A Critique of Political Economy*, Volume 1, trans. Ben Fowkes (New York: Penguin, 1982), 799.
22. Marx and Engels, *Economic and Philosophical Manuscripts of 1844 and the Communist Manifesto*, 154.

human with both technology and nature. Under capitalism, we are triply alienated: Workers do not control the products we make or the technology we use to produce them; Nature itself is increasingly privatized in a few hands; and class relations alienate humans from one another—worker against worker, consumer against producer.[23] But a truly classless society, which meets human needs through collective agency, is the only answer to such alienation. This alone can end the long-standing opposition between nature and culture, matter and thought, Gaia and Prometheus.

> Communism, as fully developed naturalism, equals humanism, and as fully developed humanism equals naturalism; it is the genuine resolution of the conflict between man and nature and between man and man [...] between the individual and the species. Communism is the riddle of history solved, and it knows itself to be this solution.

If myth is to remain relevant, it must illustrate this solution in vivid ways. We are not bound by the nightmares of the past, but free to create a new poetry of the future.[24]

---

23. Marx and Engels, *Economic and Philosophical Manuscripts of 1844 and the Communist Manifesto*, 75–78. However the fundamental form of alienation, according to Marx, is the estrangement of the human essence itself, namely, our "species being"(103).
24. Karl Marx, *The Eighteenth Brumaire of Louis Bonaparte*, trans. Daniel De Leon (New York: International, 1994), 6.

INDEX

Abrahamic religion 17, 55–56, 65–66, 71
Aeschylus 21, 47–48, 175
agonism 111, 143, 148–49
Alt-Right 98
animism 3, 108–9, 112, 125–28, 129, 131–32, 135, 140, 143, 155, 156–58, 162
Anthropocene 3
   post-Anthropocene 1
anthropocentrism 121, 129, 130, 131, 140–41
anthropomorphism 15, 66, 131–32, 154
antinomian, antinomianism 79, 82, 85, 100–1, 153, 156, 157–58, 163, 167, 175
anti-Semitism 146–48
apocalypse 3, 15–16, 39, 154
apophatism 33, 109, 113–14, 123, 139
archeofuturism 63, 67
artificial intelligence, A.I. 5, 85, 151–52, 162–63
Asherah 55–56, 61
atheism 8, 24–25, 26, 36, 72, 162
automation 4, 85, 151

Bacon, Francis 1, 26–32
Badiou, Alain 157
Bakunin, Mikhail 24–25
blaccelerationism 73–74
Brandom, Robert 82
Brassier, Ray 73–74, 76, 81–84, 87–89, 93, 100–1, 154
Brecht, Bertolt 9
Bronze Age 56, 64, 175
Brown, Dan 64–65
Burke, Edmund 96–97, 125

Campbell, Joseph 57, 60, 62–63, 68–69
Canaanite religion 55, 60–61, 63

capitalism 1–2, 9–10, 23, 38–39, 43, 48–49, 71, 92, 101–2, 176–77
   Green capitalism 49–50
   Late capitalism 43, 44, 164
carbon cycle 70–71
Christ, Carol 65–66
Chaos 5, 10, 11, 16, 23, 83, 87–88, 163
   Chaoskampf 23, 164
   Void 13, 165–66, 167
Chakrabarty, Dipesh 103, 113, 133–34, 140, 148–49, 160
Christianity 25, 30–31, 44–45, 48–49, 55–56, 65–66, 71, 106, 143
class society 6–7, 9–11, 21, 23, 38, 42–43, 46–49, 67, 176–77
climate change 1, 6, 111, 113, 127–28, 133, 135, 137, 140, 144, 174
communism 2, 46, 67–68, 72, 101, 151, 177
consumerism 3, 101, 135–36, 164
Copper Age 53–54
cosmopolitanism, cosmopolitics 118, 147, 151, 153, 175
Cronus. *See also* Chronos 12–15, 45–46
Cybernetic Culture Research Unit CCRU 74–75, 82, 95, 99
cybernetics 71, 74–75, 157
Cyborg 117, 139, 161

Danowski, Déborah 3, 103, 105–6, 107, 117, 120, 122–23, 131–39, 145–47, 157
Darwinism 71
   Social Darwinism 71–72
Deleuze, Gilles 129, 157
Demogorgon 175–76
Descartes, René 2, 70–71, 92–93, 119–20, 129

INDEX

determinism 71, 86, 90, 120–22, 129, 155–56, 170, 171–76,
   indeterminacy 86–90, 96–97, 117–18, 170
   principle of sufficient reason PSR 81
Dever, William G. 61
degrowth 1–2, 136
depopulation 1, 133
dialectics 45, 82, 113

ecology 5, 107, 118, 146
eco-modernism 1–2, 103–4
eco-pessimism, eco-pessimist 3–5, 21, 72, 103–13, 116–30, 134, 174–75
egalitarianism 57, 59–60, 64
egoism 3, 11, 34, 68
Eisler, Riane 64–65, 72
El 61
Eller, Cynthia 57–59, 63
emancipation 25, 30, 63, 73–74, 77–78, 79–80, 89, 96–97, 99, 145, 158, 173–74, 176
Engels, Friedrich 46, 67–68
enlightenment 1, 70–71, 80–82, 90–91, 96–97, 100–1, 105–8, 119, 135, 147, 163
ethnocentrism 145–46
evolution 70–71
existentialism 84–87, 91–92, 119, 167

Faust 31–35, 45–46
feminism 48–49, 133, 161
   feminist spirituality 50, 51–65, 110
   postmodern feminism 164–65
   Xenofeminism 77–78, 84, 153–54, 160
Foster, John Bellamy 1–2
Foucault, Michel 101
Frankenstein 43–47, 165–66
freedom 4, 21, 38, 40, 46, 67–68, 90–91, 141, 158–59, 172
Freud, Sigmund 9, 132
futurism 4–5

Gaia hypothesis 68–72
Gaia theory 69–71, 104, 110
Galilei, Galileo 70–71, 104–5, 129
geo-engineering 1–2, 103–4
Gimbutas, Marija 50–66, 72, 110
Globe, Globalism 105–6, 113–16, 144, 145–48
Goddess Hypothesis 54, 60–61

Goddess Movement 50–52, 57–58, 63, 68
Goethe, Johann Wolfgang 30–36, 45, 46
Gnosticism 106–7, 113–14, 119
Green Movement 53

Hallward, Peter 157
Hamilton, Clive 103, 107, 112–13, 135
Haraway, Donna 3, 103, 113–18, 128–29, 132–34, 139–43, 147–49, 152, 157, 161–62
Hardt, Michael 157
Hegel, G. W. F. 13, 82, 87, 89, 113, 116, 155–56, 160, 167–68, 169–70
Heidegger, Martin 84–85, 135
Hesiod 10–11, 15–26, 40, 41, 45–48, 62, 63, 163–64, 175
   Theogony 10, 11–12, 13, 16, 19–20, 24, 163
   Works & Days 10–11, 24
Hinduism 62–63
d'Holbach, Paul-Henri-Dietrich, Baron 108, 129
holism 70–72, 110, 139
Homer 9
homeostasis 69
human nature 6, 36–37, 85, 86, 96, 132, 134–35, 141, 146, 148–49, 174, 176
humanism 6–7, 21, 78, 87, 100, 101, 121, 125, 154–55, 177
   anti-humanism 5–6, 21, 72, 76, 86–87, 107, 130, 132, 134, 148, 151–52
   inhumanism 74, 76, 92, 100, 101–2, 154–55
   post-humanism 6
humility 1, 20, 28, 112, 135–39, 158–63, 167–68, 174
hyperstition 76, 93–100, 112, 151, 158–63, 174

ideology 4, 6–7, 9–10, 21, 37–38, 42, 45, 47–48, 49, 66–67, 72, 73–74, 75–76, 78, 84, 98, 145
Indo-Europeans 50–56
innocence, non-innocence 125, 140, 141–42, 143
instrumental reason 6, 56
Iron Age 10–11, 15–16, 21, 23, 164

James, William 118
Judaism, Jews 35, 55, 58, 106, 145, 148
Jungianism 8

INDEX

Kant, Immanuel 43–44, 88–89, 91–92, 114, 116, 122–23
Kurgans, Kurgan Hypothesis 50–51, 53–55, 60–61, 64

labor 20, 26, 34, 42–43, 48–49, 59–60, 67, 89–90, 94, 140
Laboria Cuboniks Collective 73–74
Land, Nick 73–74, 81, 99, 102
Laplace, Pierre-Simon 90, 155–56
Latour, Bruno 3, 103–11, 113–23, 126, 127, 137–38, 141, 143, 145–49, 152–54, 156–57, 159, 160, 171
Leibniz, Gottfried 120, 123
localism 73–74, 97, 124–25, 136, 146
  locavorism 3
logocentrism 128
Lovecraft, H. P. 99, 165–66, 167–68
Lovelock, James 68–72, 104–5, 110, 114–15, 122

magic 32, 33, 52, 162, 171
de Maistre, Joseph 96–97, 125
Malm, Andreas 2
Marduk 17, 164–65
Marxism 2, 6–7, 10, 21, 43, 46, 59–60, 67–68, 72, 76–77, 82, 87, 106–7, 108, 120, 134–35, 176
materialism 108, 118–19, 120, 129–30, 153–54, 155–56, 162, 171
matriarchy 55, 57, 59–61, 63, 66–67
Matrifocal Society 55, 56–57, 58–59, 61, 62, 64, 65–67
Matrilineal Society 55–56, 58–59, 65
matter 24, 43, 81–82, 85–87, 94–95, 106–7, 108–9, 118–20, 121, 128, 135, 152, 153–54, 156, 176–77
mechanism, mechanistic 2, 70–72, 85, 108–10, 126–29, 137, 153–54, 172–73
Midgley, Mary 70–72
misanthropy 1, 6, 41, 46
monad 120, 123
monism 129, 173–74
monotheism 55–56, 58, 60–61, 147
Moore, Jason 2
Morton, Timothy 138–39, 157
moses, mosaic 146–49

multinaturalism 122
multiplicity 74, 110–11, 113–14, 124, 132, 138, 141, 156, 157
mystery, mysticism 5, 7–10, 23, 52, 57–58, 72, 86–87, 99, 125, 126, 131–32, 138–39, 147, 155, 162, 170, 175–76
  versus the "Mythic" 23, 176–14

natural law 7, 80–82, 87–88, 94–95, 110–11, 130, 153–54, 160, 162–63, 171
naturalism 6–7, 21, 77, 89, 111, 129–30, 177
Nazism 38, 143, 145, 148
Negarestani, Reza 73–74, 82, 86–89, 92, 100, 153, 154–55, 171
negative theology 114, 123, 139, 162
Negri, Antonio 157
Neolithic Age 51–54, 56–57, 59–60
neoreactionary 43
Nietzsche, Friedrich 34–40, 45–46,
nominalism 6, 36, 76, 77–86, 87–88, 94–95, 112, 154–56, 162
noumenalism 112–16, 119–20, 122–24, 126, 155, 162
  black box 114–15, 119–20, 122–23, 126, 162

objectivity 2, 77, 78, 84, 87–93, 106, 120, 123–25, 130, 138–41, 144, 147, 152, 158–63, 174–76,
Object-Oriented Ontology 109, 119
object-oriented politics 107
Oikeiôsis 172
Old Europe 51–57, 59–61, 63
Olympian 14–17, 24, 30–31, 46–47, 163, 175–76
O'Sullivan, Simon 99

Pagan, Neo-Pagan 65–66
Pandora 19–20
Pan-psychism 129–30
Parisi, Luciana 85, 152
perspective, perspectivalism 116, 122–23, 138, 141–45
Peterson, Jordan 98, 164–65,
phenomenology 82–84, 120–21, 128–38, 152–56, 162
Phillips, Leigh 3–4
Plato 9, 116, 123–24, 172–73

pluralism, plurality 85, 94–95, 113–14, 117–18, 122–23, 129–30, 132, 139, 143, 146
pluriverse 111, 117–18, 145, 146, 159
postmodernism 8, 72, 106, 139, 157, 164–65, 167
premodern 42–43, 59–60, 67, 128, 135

Rand, Ayn 38–46
rationalism 9, 72, 80–82, 86–87, 90–92, 96–97, 106–8, 110–11, 113–14, 118, 120–22, 143, 158, 167–69, 172–73, 176
reductionism 72, 75, 116–17, 125, 130, 141
Reed, Patricia 89–90, 95
religion 7–8, 36, 52, 53, 55–57, 58, 59, 60, 61, 65–67, 71, 121–22, 137, 144
romanticism 5, 96–97
Ruether, Rosemary Radford 65

Sartre, Jean-Paul 87, 91–92, 148
Satan 24–25, 165
Schmitt, Carl 143, 144–45, 146–47, 148–49, 159, 160
scientism 100, 154
Sellars, Wilfrid 82–83, 88–89
Shelley, Mary 26, 43–45, 46–47, 165–66
Shelley, Percy Bysshe 21, 175
Singleton, Benedict 79–80
skepticism 103, 168–69
socialism 1–2, 26, 40–41, 48, 72, 101
Socrates 8–9
Spinoza, Baruch 27–81, 89, 90–91, 97, 108, 129, 155–56, 168, 169, 170, 172
Srnicek, Nick 95–96

Stengers, Isabelle 3, 103, 105–6, 111, 113–14, 118, 123–25, 134–35, 138–39, 148–49, 160–61
Stone, Merlin 55–56
substance 80–82, 109, 123–24, 163, 167, 170–73, 176

Tiamat 14, 17, 164–65
Titanomachy 14, 112
Trans Exclusionary Radical Feminism TERF 78
Trump, Donald 97–98
Typhon 14–15, 16, 17, 20

universal basic income UBI 4
Uranus 11–12, 13, 14–15, 16, 45–46
utopia, utopianism 2, 42, 43, 53–54, 57, 67, 92, 96, 98–99, 165

Viveiros de Castro, Eduardo 3, 103, 105–6
Voegelin, Eric 106–7
voluntarism 76–77, 87–93, 94–95, 112, 156–57, 159, 162

Wark, McKenzie 104, 122
white nationalism 73–74, 98
Williams, Alex 95–96
Wolfendale, Peter 76, 92, 157

Yahweh 14–15, 26, 60–61

Zeus 13–15, 16–20, 24, 25, 27, 28, 30–31, 32, 36, 37, 46–48, 112, 163, 164–65, 166, 175
Žižek, Slavoj 157

www.ingramcontent.com/pod-product-compliance
Lightning Source LLC
Chambersburg PA
CBHW021142230426
43667CB00005B/225